Advances in Anatomy Embryology and Cell Biology

Vol. 167

Springer-Verlag
Berlin Heidelberg
GmbH

M. von Lüdinghausen

The Clinical Anatomy of Coronary Arteries

With 27 Figures and 3 Tables

Springer

Prof. Dr. MICHAEL VON LÜDINGHAUSEN

Institut für Anatomie und Zellbiologie,
Universität Würzburg,
Koellikerstr. 6,
97070 Würzburg, Germany
e-mail: anat033@mail.uni-wuerzburg.deauthorinfo

ISSN 0301-5556
ISBN 978-3-540-43689-8

Library of Congress-Cataloging-in-Publication-Data
The Clinical Anatomy of Coronary Arteries / M. v. Lüdinghausen - Berlin; Heidelberg; New York: Springer, 2003
(Advances in anatomy, embryology, and cell biology; Vol. 167)
ISBN 978-3-540-43689-8 ISBN 978-3-642-55807-8 (eBook)
DOI 10.1007/978-3-642-55807-8

http://www.springer.de

Originally published by Springer-Verlag Berlin Heidelberg New York in 2003

Production: PRO EDIT GmbH, 69126 Heidelberg, Germany
Printed on acid-free paper - SPIN: 10993798 27/3111 - 5 4 3 2 1

Contents

1 Introduction

Considerable advances have been made in cardiology during the last few decades. In particular, there has been great progress in the field of coronary angiography both when combined with, and without, computed tomography (CT) and magnetic resonance (MR) imaging. These techniques of modern imaging allow the cardiologist and coronary surgeon to study every cardiac structure in detail, both two- and three-dimensionally and from either side, to analyze the movements of the heart and valves, and to observe myocardial circulation and even myocardial metabolic processes.

However, coronary heart disease, a multifactorial illness of the coronary vessels, still remains the most common cause of death in developed countries.

In addition to the large group of patients suffering from coronary heart disease, there is a smaller group of children and adults who are in need of open heart surgery and, most frequently, valve surgery. A very small number of individuals suffering from Wolff-Parkinson-White syndrome still await competent surgical intervention.

These three groups of patients have in common that, for them, meticulous preoperative diagnostics and preparation for surgery are urgently required. Any open heart surgeon who carries out procedures in the coronary or interventricular grooves or on the atrial walls of the heart must take the normal and anomalous origins, courses, and terminations of cardiac vessels into consideration. Therefore, with the availability of precise anatomical and physiological data, operation time will be shortened, operative risks will diminish, and the safety of the operation for the patient will be greater.

Morbidity and mortality rates among our patients over the last five decades oblige us to extend the frontiers of understanding where both the macro- and microanatomy and also the pathophysiology of heart diseases are concerned, and develop new approaches for the diagnosis and treatment of myocardial dysfunction, which is largely a consequence of variations in the morphology of the coronary arteries and cardiac veins.

The results of 25 years of study of the intrinsic vessels of the human heart are presented in this work; the findings have been described and evaluated from the viewpoint of a clinically oriented anatomist, and have been compared with, and discussed in the light of the results of other research in the field as mentioned in the international literature.

1.1
Aims of the Study

The main objectives of this work are to illustrate the most recent research into the macro- and micro-anatomical morphology of the coronary arteries, to discuss controversies surrounding the anatomical basis for arterial catheterization, and to establish the anatomical prerequisites necessary before minimally invasive and surgical interventions can be carried out. Areas which merit further study are identified.

The issue of myocardial microcirculation is not considered in this work.

1.2
The Cardiac Vascular Systems: The Coronary Arteries and Cardiac Veins

The myocardial walls of the four-chambered heart are nourished by both the right and left coronary arteries. The venous drainage is guaranteed by the cardiac veins, i.e., partly by tributaries of the CS and partly by tributaries of cardiac veins from the right

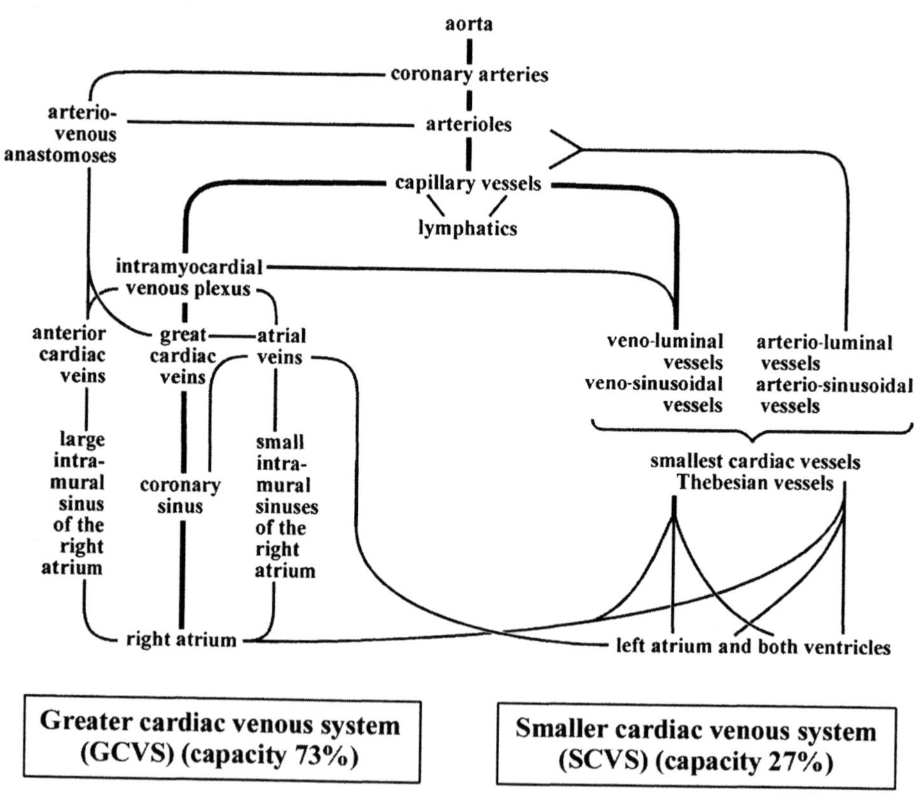

Fig. 1. Schematic drawing to illustrate the distribution pattern and interconnections of the tributaries of the greater (*GCVS*) and smaller cardiac venous systems (*SCVS*)

and left atrium (Schlesinger 1940a; Adachi 1928; Mochizuki 1933; Cranicianu 1922; Hadziselimovic 1982).

Additionally, there are a few noncoronary extracardiac arteries and veins (Halpern and May 1958; Petelenz 1965b; Bloor and Liebow 1965; Moberg 1967a, b). All these vessels join the "greater or major cardiac venous system" (GCVS).

The "smaller or minor cardiac venous system" (SCVS) comprises the smallest cardiac (formerly Thebesian) vessels in the innermost layers of all cardiac chambers, consisting of arterio-luminal, arterio-sinusoidal, veno-luminal, and veno-sinusoidal vessels (Hadziselimovic 1982; Gould 1953; Cornel et al. 1992; von Lüdinghausen et al. 1995) (Figs. 1, 2).

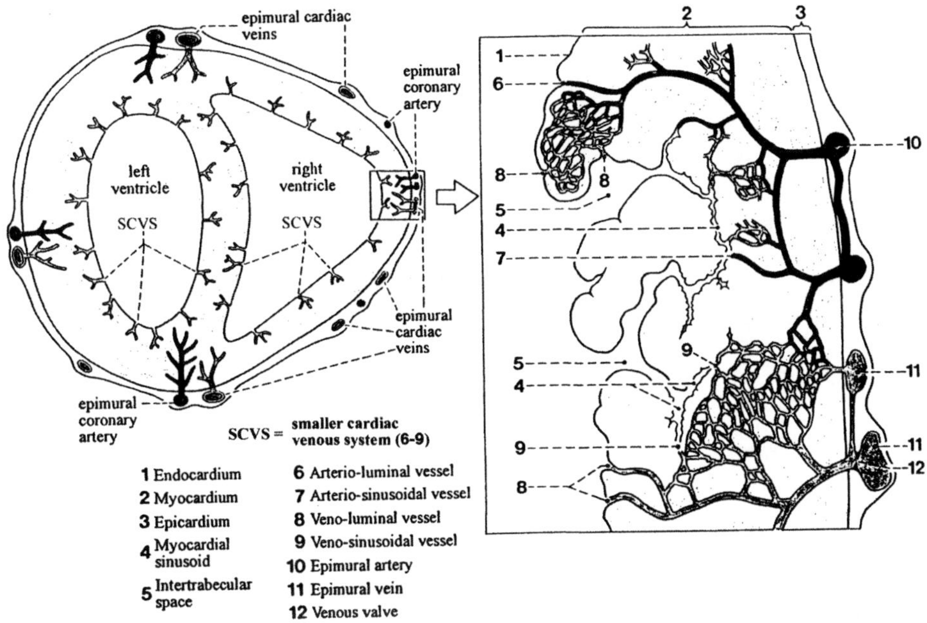

Fig. 2. Distribution pattern of the cardiac vessels in the myocardium of the ventricular walls. Schematic drawing of a cross-section through the ventricles showing the vessels of the GCVS on the myocardial surface and those of the SCVS in the subendocardial layers. Enlargement of a section of the left ventricle (indicated by a *rectangle*) showing the arrangement of the epimurally distributed coronary arteries and cardiac veins, the intramural distribution of their branches and their relation to subendocardial vessels

2 The Significance of the Proper Vessels of the Human Heart in Statistics and Diagnostics

Coronary heart disease constitutes one of the major health care problems of the Western Hemisphere: this widespread and life-threatening condition afflicts more than 5 million people in the United States (National Institute of Health 1993). It causes one-third to one-half of all deaths of people between the ages of 35 and 64 years. Up to 13% of middle-aged men in the general population have coronary artery disease, many without symptoms. Consequently, 40%–60% of patients who suffer myocardial infarction or die suddenly present without prodromi (Gould 1988).

In European countries, coronary heart disease afflicts more than 10 million people (European Atherosclerosis Study Group 1988) and is also the leading cause of death, killing more than 1 million individuals annually (Effler and Sheldon 1974; Hadziselimovic 1982; Paschal et al. 1993). Calcified deposits in plaques within coronary arteries have been shown to be an indicator of coronary artery disease (Mautner et al. 1994).

In recent years, there has been rapid development and refinement of new techniques in the diagnosis and treatment of coronary heart disease. More precise assessment of the extent of coronary atherosclerosis and its calcifications is possible with improved angiography and high frequency epicardial echocardiography. Innovations such as electron beam CT, percutaneous transluminal balloon angioplasty (PTCA), laser angioplasty, coronary bypass graft surgery and placement of coronary artery stents are being used more widely (Giraldo and Higgins 1988; Mautner et al. 1994).

Currently, in the United States alone, some 900,000 patients per year are subjected to selective coronary angiography, and more than 300,000 per year undergo coronary angioplasty or bypass surgery (Angelini 1989; Jones et al. 1994).

Consequently, great emphasis should be placed on the clinical significance of the proper myocardial vessels not only with regard to middle-aged individuals, but also septuagenarians and octogenarians; this is the group of patients which our medical students, the future physicians, will one day be called upon to treat.

For the measurement of ventricular mass and the visualization of the anatomy of the coronary arteries the following procedures are widely recommended: two-dimensional *echocardiography* (Kaul et al. 1986; Clausen et al. 1988; Douglas et al. 1988; Kaul 1989; Byrd et al. 1989; Cornel et al. 1992; Riopel 1993; Macchi et al. 1994) and *ultrasonography* (De Brux et al. 1987).

Nuclear cardiological techniques are used for the evaluation of myocardial perfusion, perfusion reserve and myocardial revascularization:

1. Myocardial scintigraphy, using positron emission tomography (PET) (Marshall et al. 1983; Corbett et al. 1987; Positron Emission Tomography Panel 1988; Gould et 1986; Schelbert 1994)
2. Ultrafast thallium-201 single photon emission computed tomography (SPECT) (Clausen 1988; Johnston and Pohost 1994; Brundage 1994)

For the assessment of cardiac vessels, especially where the origin and distribution of coronary arteries is concerned, the following techniques are widely recommended: selective coronary (cine-)*angiography* (Paulsen and Vetner 1973; Kaltenbach and Becker 1974; Vlodaver et al. 1976; Arnett et al. 1979), *magnetic resonance imaging* (MR imaging) (Akins et al. 1985; Müller et al. 1997), *ultrafast magnetic resonance imaging* (ultrafast MR imaging) (Atkinson et al.1990), *cinemagnetic resonance imaging* (cine-MR imaging) (Aurigemma et al. 1989), *magnetic resonance angiography* (MR angiography) (Alfidi et al. 1987; Pettigrew 1994; Pennell et al. 1996; Sandstede et al. 1998; Kessler et al. 1999; Knez et al. 2000), conventional *computed tomography*, ultrafast computed tomography, contrast-enhanced electron-beam computed tomography, and electrocardiographically triggered electron beam computed tomography (Brundage 1994; Haberl et al. 1998, Knez et al. 2000; Schaffler et al. 2000).

2.1 Angiography

Among the above-mentioned techniques, selective coronary cineangiography through coronary artery catheterization (Conti et al. 1973; Kimbiris et al. 1978; Paulin 1964, 1983; Atkinson et al. 1990; Day et al. 1992; Paschal et al. 1993; Riopel 1993) is at present the method of choice or the "gold standard" technique for determining the many variations and anomalies of the cardiac vessels and for ascertaining the morphological changes due to coronary arteriosclerosis (location of the obstruction and the presence, location, and degree of coronary arterial narrowing), in patients exhibiting stable or unstable angina pectoris (so-called preinfarction angina) and in patients with complete myocardial infarction.

The double contrast technique enhances the results of postmortem coronary angiography, especially in the radiological study of coronary ostia (Rissanen 1970; Köhler et al. 1981).

No overestimations of the degree of luminal narrowing by atherosclerotic plaques were made in studies carried out by Arnett et al. (1979); the results were checked histologically. Only a few cases were underestimated.

The results of coronary (cine-)angiography generally form the basis for many decisions concerning the therapy of patients with symptomatic coronary narrowing (Conti et al. 1973; Vlodaver et al. 1976; Conti 1977). In fact, the technique provides valuable information, and at lower cost, but it is invasive and involves risks associated with the use of catheters, contrast agents, and ionizing radiation. As stated above, a few narrowings have occasionally been missed by selective cineangiography (Arnett et al. 1979).

Postmortem coronary angiograms have never been found to show the degree of narrowing of the coronary arteries accurately; their findings have been underestimated to a considerable degree (in up to one-third of all cases according to Vlodaver et al. 1973, Grondin et al. 1974, Schwartz et al. 1975, and Arnett et al. 1979).

It may also be difficult to recognize the various coronary anomalies angiographically; failure to do so may prolong the angiographic procedure or lead to repeat catheterizations (Donaldson and Raphael 1982; Click et al. 1988; Kessler et al. 1999).

2.2 MR Imaging

MR imaging is a noninvasive investigative technique with vast potential for defining cardiac anatomy and function. It has become the preferred method for the topographical visualization of the left heart using transverse, coronal, and sagittal planes and for noninvasive screening of ischemic heart disease with patients in a supine position (Akins et al. 1985). The method can be used to study occlusive lesions in the perfused heart of a human subject without known cardiac disease. Optimally orientated images provide useful information about cardiac morphology and motion. The accuracy of the technique and the reproducibility of the images obtained are well established. It does not carry the risks associated with the use of catheter and contrast agent, does not involve ionizing radiation, and does not cause physical discomfort to the patient (Crooks et al. 1984; Kaul et al. 1986; Katz et al. 1988; Aurigemma 1989; Atkinson et al. 1990; Gates et al. 1994).

2.3 Ultrafast MR Imaging

Ultrafast MR imaging has the advantage that it can provide adequate temporal and spatial resolution to permit first-pass perfusion studies of the heart (Atkinson et al. 1990; Bormann et al.1992).

Cine-MR imaging is well suited for the visualization of normal vascular and intracardiac blood flow as well as flow disturbances associated with cardiac disease. This technique is potentially useful in detecting and gauging the severity of mitral and aortic insufficiency, for evaluating left ventricular function, and for detecting and determining coronary bypass graft patency and, potentially, coronary artery stenoses. Unlike cine-computed tomography, neither contrast injection nor radiographic exposure is required. Furthermore, with recent software advances, good quality coronal, sagittal, and oblique imaging can readily be carried out without patient repositioning (Aurigemma et al. 1989).

2.4 Coronary MR Angiography

Coronary MR angiography constitutes a noninvasive technology that could be used in outpatient surveys of populations at risk (Alfidi et al. 1987; Dumoulin et al. 1991; Anderson et al. 1993; Paschal et al. 1993). Moreover, three-dimensional reconstructions of cardiac vessels enable the elucidation of even very rare, unexpected findings.

MR imaging of the coronary arteries to establish their individual distribution patterns, localize stenosis and measure its severity, and assess the distal vessel is important in the management of coronary artery disease, and is essential before revascularization by coronary bypass surgery or angioplasty (Cho et al. 1991; Pennell et al. 1996).

Three-dimensional coronary MR angiography with navigator techniques and MR angiography with fast contrast-enhanced three-dimensional breath-hold imaging allow the almost brilliant imaging of the proximal course of anomalous coronary arteries, especially visualization of most significant coronary stenoses (Sandstede et al. 1998; Kessler et al. 1999).

When the many diagnostic, noninvasive, and invasive procedures are used there should be an awareness that their results and interpretations are very much influenced by the cardiologist's or coronary surgeon's experience and knowledge of the anatomy of the heart, and especially its proper vessels, in each individual patient (Blumgart 1940; James 1961; Blunk and DiDio 1971; Allwork 1987; Anderson et al. 1993; Pennell et al. 1996).

Technical progress is being made at a rapid rate. Given the fact that, for instance, cine-MR imaging requires neither radiation nor intravenous contrast injection, further development appears to be warranted. However, any technical progress is worthless and contributes nothing to the benefit of patients if detailed consideration is not given to the large number of coarse or tiny, frequent or rare variations in individual anatomy.

2.5 Conventional and Ultrafast Computed Tomography (Cardio-CT)

Cardio-CT provides sufficient morphological information in the case of congenital heart disease, volume or pressure loading, left ventricular aneurysm, tumors, thrombi, cardiac calcifications, and pericardial disease. It allows qualitative assessments of myocardial thickness and of the perfusion effected through a coronary bypass (Lackner et al. 1988; Fallavollita et al 1994).

Efficient or even good evaluation of the heart by conventional CT has been severely limited because of blurred pictures created by cardiac contraction and relaxation.

Ultrafast CT avoids these disadvantages; the greatest clinical experience gained with its use has been in the measurement of myocardial mass and in the evaluation and study of coronary artery disease. The excellent spatial and temporal resolution of ultrafast CT makes it an ideal method for the evaluation of left ventricular aneurysm and associated thrombus (Brundage 1994).

Electrocardiographically triggered, contrast-agent enhanced, electron beam computed tomography is used to investigate a wide range of cardiac diseases, especially coronary heart disease with and without calcification deposits. It is possible to distinguish quite clearly between the coronary arteries, cardiac veins, and the CS (Mautner et al. 1994; Schaffler et al. 2000).

3 Nomenclature

In this work, the accepted terminology of the branches of the left coronary artery (LCA) and right coronary artery (RCA) is used in its English and Latin versions according to Forssmann (1979), Nomina Anatomica (1980), and Terminologia Anatomica (1998), with angiographical nomenclature according to Kaltenbach and Spahn (1975).

3.1 English and Latin Versions

The English names are given in italics, the Latin names are given in parentheses/brackets. The numbers correspond with those of Fig. 3.

Left Coronary Artery (A. coronaria sinistra) and Its Branches (Fig. 3)

1. Main stem of the left coronary artery (Truncus proximalis ae. coronariae sinistrae)

Ventricular Branches

2. Anterior interventricular branch (artery) (R. interventricularis anterior)
3. (Left) conus branch [R. (sinister) coni arteriosi]
4. Left superior septal branch (R. superior sinister septi interventricularis)
5. Anterior interventricular septal branches (Rr. interventriculares septales anteriores)
6. Anterior diagonal branches; lateral diagonal branches; lateral branches (Rr. diagonales anteriores et laterales, Rr. laterales)
7. Anterior branches of the right ventricle (Rr. anteriores ventriculi dextri)
8. Median or intermedian branch(es) [R. medianus; R. intermedius; variation(s) in a case of trifurcation or quatrofurcation of the main stem]
9. Circumflex artery (branch) (R. circumflexus)

Ventricular Branches

10. Lateral diagonal branches (Rr. diagonales laterales)
11. Left marginal branch (R. marginalis sinister)
12. Posterior branch of the left ventricle (R. posterior ventriculi sinistri)
13. (Left) atrioventricular branches (Rr. atrioventriculares sinistri)

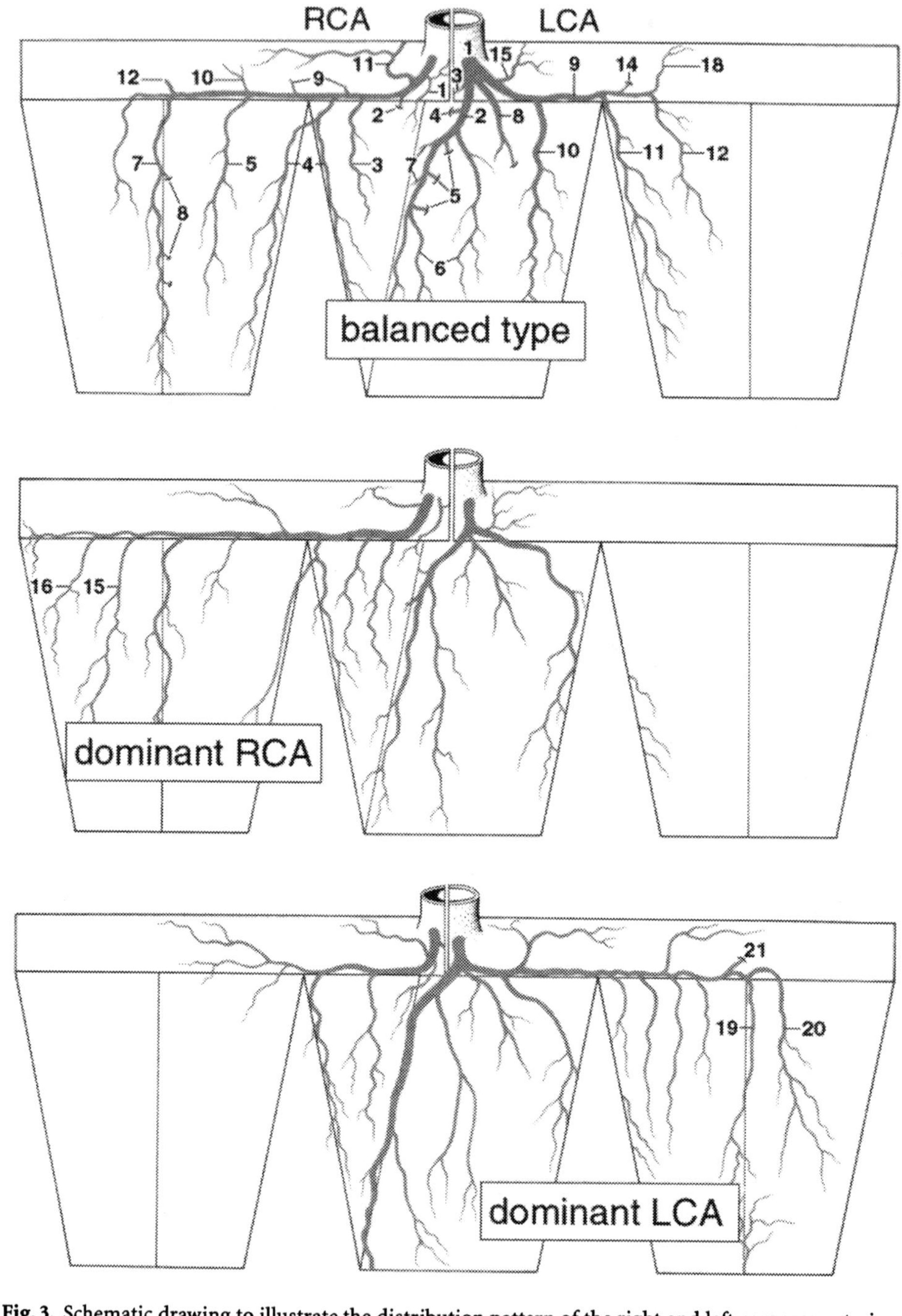

Fig. 3. Schematic drawing to illustrate the distribution pattern of the right and left coronary arteries (*RCA*, *LCA*) according to coronary dominance; the balanced type, the dominant RCA and the dominant LCA are shown

Atrial Branches

14. Atrial anastomotic branch (Ramus atrialis anastomoticus)
15. (Left) sinuatrial node branch [R. nodi sinuatrialis (sinister) (variation)]
16. Left lateral atrial branches (Rr. atriales sinistri laterales)
17. Left intermediate atrial branch (R. atrialis sinister intermedius)
18. Left posterior atrial branches (Rr. atriales sinistri posteriores)

Left coronary artery:

1. Main stem of the left coronary artery
2. Anterior interventricular artery (branch)
3. (Left) conus branch
4. Left superior septal artery (branch)
5. Anterior septal branches
6. Anterior diagonal branches; antero-lateral branches
7. Anterior branches of the right ventricle
8. Median or intermedian branch(es), variation(s) in a case of trifurcation or quadrufurcation of the main stem
9. Circumflex artery (branch)
10. Lateral diagonal branches
11. Left marginal branch
12. Posterior branch of the left ventricle
13. (Left) atrioventricular branches
14. Atrial anastomotic branch
15. (Left) sinuatrial node branch
16. Left lateral atrial branches
17. Left intermediate atrial branch
18. Left posterior atrial branches

Derivatives of a dominant LCA:
19. Posterior interventricular artery (branch)
20. Postero-lateral branch of the right ventricle
21. (Left) atrioventricular node branch (variation),
(Branches 13, 16, and 17 mentioned above are not shown in the figure)

Right coronary artery:

1. (Right) conus branch
2. Right superior septal artery
3. Right anterior ventricular arteries
4. Right marginal branch
5. Right posterior ventricular branches
6. Right posterior atrioventricular branches
7. Posterior interventricular artery (branch)
8. Posterior septal branches
9. Right anterior atrial branches
10. Right intermediate atrial branch
11. (Right) sinuatrial node branch
12. (Right) atrioventricular node branch
13. Atrial anastomotic branch
14. Right posterior atrial branch

Branches running from a dominant RCA:
15. (Right) posterior branch(es) of the left ventricle
16. Posterolateral branch of the left ventricle
(Branches 6, 13, and 14 mentioned above are not shown in the figure)

Derivatives of a Dominant LCA

19. Posterior interventricular branch (artery) (R. interventricularis posterior)
20. Postero-lateral branch of the right ventricle (R. postero-lateralis ventriculi dextri)
21. (Left) atrioventricular node branch (R. nodi atrioventricularis sinister) (variation)

A. coronaria dextra (Right Coronary Artery) and Its Branches (Fig. 3)

Ventricular Branches

1. (Right) conus branch (R. (dexter) coni arteriosi)
2. Right superior septal branch (R. superior septi interventricularis dexter)
3. Right anterior ventricular branches (Rr. anteriores ventriculi dextri)
4. Right marginal artery (R. marginalis dexter)
5. Right posterior ventricular branches (Rr. posteriores ventriculi dextri)
6. Right posterior atrioventricular branches (Rr. atrioventriculares posteriores dextri)
7. Posterior interventricular artery (branch) (R. interventricularis posterior)
8. Posterior interventricular septal branches (Rr. interventriculares septales posteriores)

Atrial Branches

9. Right anterior atrial branches (Rr. atriales dextri anteriores)
10. Right intermediate atrial branch (R. atrialis dexter intermedius)
11. (Right) sinuatrial node branch [R. (dexter) nodi sinuatrialis, A. nodi sinuatrialis (dexter)]
12. (Right) atrioventricular node branch [R. nodi atrioventricularis (dexter)]
13. Atrial anastomotic branch (R. atrialis anastomoticus)
14. Right posterior atrial branch (R. atrialis dexter posterior)

Derivatives of a Dominant RCA

15. Posterior branch of the left ventricle (R. posterior ventriculi sinistri)
16. Posterolateral branch of the left ventricle (R. posterolateralis ventriculi sinistri)

3.2 Abbreviations

The anterior and posterior interventricular and circumflex vessels and the main arteries for the conduction system are designated *arteries*, and their further minor ramifications as *branches*, according to the suggestions and proposals made in the Anglo-American literature (Baroldi and Scomazzoni 1965; McAlpine 1975; Angelini 1989).

AIA	Anterior interventricular artery
AO	Aorta
ATH	Atheromatous plaque
AV node	Atrioventricular node
AV node branch or artery	Atrioventricular node branch or artery
CA	Circumflex artery
CAL	Calcification
CS	Coronary sinus
GCV	Great cardiac vein
IVC	Inferior vena cava
IV septum	Interventricular septum
IV sulcus	Interventricular sulcus
LA	Left atrium
LCA	Left coronary artery
LDSA	Left descending septal artery
LSSA	Left superior septal artery
LV	Left ventricle
MI	Mitral valve
PIA	Posterior interventricular artery
PIV	Posterior interventricular vein
PT	Pulmonary trunk
PVV	Pulmonary veins
RA	Right atrium
RV	Right ventricle
RCA	Right coronary artery
RSSA	Right superior septal artery
SA node	Sinuatrial node
SA node branch or artery	Sinuatrial node branch or artery
SCV	Small cardiac vein
TRI	Tricuspidal valve

3.3 Commonly Used Unofficial or Alternative Terms and Synonyms

Alternative names commonly used in the clinics and other synonyms for the superficial coronary arteries are listed here.

- **Adventitial branches:** Vasa vasorum (McAlpine 1975).
- **Anterior descending artery:** Anterior interventricular artery (branch).
- **Conus artery:** Conus branch, infundibular artery (Becker 1995).
- **Crista supravalvularis aortae:** Supravalvular crest, supravalvular ridge, sinus rim, aortic bar.
- **Crux cordis:** Situation of the heart posteriorly, where the auricles and ventricles meet (Gross 1921); the conjunction of the coronary, interventricular and interatrial sulci (James 1961; Williams et al. 1989), intersection of sulci at the diaphragmatic aspect of the heart (DiDio and Wakefield 1975, Paulin 1983, Arid et al. 2000).

- **Circle of Vieussens:** Subepicardially distributed anastomosis between the conus branches of the AIA and RCA.
- **Innominate artery:** Truncus brachiocephalicus, brachiocephalic trunk (Davies and Lie 1977).
- **Internal mammary artery:** A. mammaria interna; A. thoracica interna; internal thoracic artery.
- **Koch's triangle:** Area between the attachment of the septal or medial leaflet of the tricuspid valve, the antero-medial margin of the coronary sinus orifice, and the round collagenous, subendocardial tendon of Todaro at the right fibrous trigone or central fibrous body (Williams et al. 1989; Arid et al. 2000).
- **Kugel's artery (Kugel 1927):** R. atrialis anastomoticus: arteria anastomotica auricularis magna, atrial anastomotic branch, variable anastomosis in the lower portion of the atrial septum connecting the CA and the RCA.
- **Left anterior descending artery, left descending artery:** anterior interventricular artery, R. interventricular anterior.
- **Left anterior ventricular arteries:** Rr. anteriores, Rr diagonales, Rr anterolaterales.
- **Left circumflex artery (Paulin 1983):** Circumflex branch (artery).
- **Left margin:** Margo sinister, formerly obtuse margin, margo obtusus.
- **Left (anterior) aortic sinus:** Left posterior sinus of the ascending aorta (Williams et al. 1989).
- **Moderator band artery (Reig et al. 2000), descending septal artery:** R. descendens of the anterior interventricular artery, i.e. the 2nd septal branch of the AIA.
- **Non-coronary aortic sinus:** posterior aortic sinus, right posterior sinus of the ascending aorta.
- **Preventricular branch (Barry and Patten 1953; McAlpine 1975):** R. anterior ventriculi dextri; right anterior ventricular artery.
- **Pulmonary cone (Barry and Patten 1953):** Conus arteriosus, infundibulum of right ventricle.
- **Ramus ostii venae cavae superioris (Paulin 1983):** sinuatrial node artery, sinus node artery.
- **Right (anterior) aortic sinus:** Anterior aortic sinus of the ascending aorta (Williams et al. 1989).
- **Right circumflex artery:** The proximal RCA forming a semicircle in the right coronary sulcus.
- **Right margin:** Margo dexter, formerly acute margin, margo acutus.
- **R. obliquus:** R. medianus, R. diagonalis, oblique branch, median branch, diagonal branch.
- **Sinuatrial node branch or artery:** Sinuauricular node artery, sinus node artery (James 1961), ramus cristae terminalis (Spalteholz 1924).
- **(Right, left) superior septal artery:** Superior septal branch, first septal perforator coronary artery (Rath et al. 1986), right superior septal perforator artery (Bream et al. 1979), R. superior ventriculi septi, branch or artery of crista supra-ventricularis (McAlpine 1975, Campbell 1929), descending septal branch or artery (Rodriguez et al. 1961, Taylor 1980).
- **Sinus of Valsalva:** Sinus aorticus, aortic sinus.
- **Third primary division (Pepler and Meyer 1960):** median, intermedian or diagonal branch of the main stem of the LCA.
- Vieussens' circle (see above: "Circle of Vieussens").

4 Results and Discussion

4.1
Anatomy of the Coronary Arteries in General

Development. In the early embryonic life of the human, before the coronary arteries appear, the heart wall is a loose meshwork of developing myocardial fibers, with relatively large spaces between the muscle trabeculae containing blood which circulates back and forth from the cavities with the heartbeat. These spaces or sinusoids enable the supply of the myocardium (Licata 1955, 1956; O'Rahilly 1971; Bankl 1977; van Mierop 1979).

In the seventh embryonic week, each coronary artery develops, initially, from two anlagen. That which is distal and near to the heart chambers is the forerunner of the subepicardial branches of the coronary arteries; the other, proximal and near to the truncus arteriosus is known as the peritruncal ring (Bogers et al. 1989). The distal anlage develops as a subepicardial vascular network in the coronary and interventricular sulci and on the walls of the ventricles and bulbus. These networks – the anlagen of the right posterior and left anterior arteries – rapidly extend over the heart (Hirakow 1983; Hutchins et al. 1988; Conte and Pellegrini 1984; Bogers et al. 1989; Christ 1990).

The proximal anlage, which is the forerunner of the stems of the right and of the left coronary arteries, develops as several buds on the walls of the truncus arteriosus. Normally, only two endothelial buds (or sprouts, right and left), hollow out, increase in length and connect (at Carnegie stage 18, according to Hutchinson et al. 1988) with the vessels of the right and the left networks respectively, so that the coronary arteries are formed. This is in agreement with the reports by Licata (1955, 1956), Gregg and Fisher (1963), Doty et al. (1976), and Christ (1990) concerning humans, and by Dbaly et al. (1968) and Voboril and Schiebler (1969) concerning rats.

The coronary system is, to all purposes, completed in the 45-mm stage (Robb 1965). Some of the vessels connect with the endothelium-lined sinusoids; therefore, some communications (continue to) exist within the cardiac chambers. If such intertrabecular spaces and sinusoids – communicating with a ventricular lumen – persist and enlarge, fistulous connections appear (Dusek et al. 1975) (see Sect. 4.2.5).

Origin of Coronary Arteries. The epicardial coronary artery system consists of the left and right coronary arteries, which normally arise from ostia located in the left and right aortic sinuses, respectively.

Epi- and Intramural Distribution of Coronary Arteries. The initial segments of the coronary arteries are distributed epimurally and can, in most cases, easily be seen beneath the epicardium (Gross 1921; McAlpine 1975; Angelini 1989).

In contrast to the epimurally located large coronary arteries and their branches, the intramural or intramyocardial portion of the coronary artery supply consists of numerous small branches that pierce the wall in a more or less perpendicular fashion.

Small Intramural and Subendocardial Arteries and Arterioles. According to Baroldi and Scomazzoni (1965), the intramural branches generally penetrate the ventricular wall and thereafter exhibit a finer dichotomous arborization into smaller branches, which ramify in multiple planes parallel to the internal surface of the wall (Gross 1921; James 1961; Hadziselimovic 1982; Paulin 1983). Finally, they reach the arteriolar level where they terminate in the intramural capillary network, and subendocardially in a vascular plexus (Estes et al. 1966a,b; Lametschwandtner et al. 1990).

Intramural Ramifications. According to Paulin (1983), there are further ramifications into the capillary network from intramural arteries in all directions and at all levels. Distinctions can be made between intramural branches that, after a short distance, immediately subdivide into a mesh-like pattern, preferentially supplying the outer layers, and those that are longer and penetrate more deeply to supply the more endocardially located layers, including the subendocardial zone. Paulin (1983) emphasized that the subendocardially located contracting myocardium exposes the corresponding capillary networks to a phasic compression, which causes systolic reduction of coronary blood flow as well as the increased vulnerability to vascular insult of the subendocardial layers of the myocardium.

Capillary Network and Further Communications. The capillary network in the human ventricular myocardium in man consists of two sphincter-type capillaries (the met-arteriole and pre-capillary) and a nonmuscular capillary, the classic or true capillary (Provenza and Scherlis 1959).

Occasionally, a few arterioles will communicate with a cardiac cavity (right and left ventricles and right atrium), producing arterio-luminal and arterio-sinusoidal vessels and thus a diminutive coronary cameral, coronary luminal, or coronary sinusoidal connection (Gould 1953; Gregg and Fisher 1963; Paulin 1983).

How many of these are necessary in the "normal" human heart is not yet clear (Baroldi and Scomazzoni 1965; Irino et al. 1982; Angelini 1989) (Figs. 1, 2).

Generally, atrial branches supply the atrial myocardium and ventricular branches supply the walls of the ventricles. However, there are two exceptions: (1) the atrioventricular branches (DiDio and Wakefield 1972), and (2) the terminal AV node branch, which also supplies the AV bundle and proximal right and left bundle branch (James 1961).

Histology of Coronary Arteries. The coronary arteries are arteries of the fourth order, their caliber corresponding to that of the distal portion of the radial arteries. The coronary arteries are of muscular type, showing a sudden transition from the aorta; they are characterized by the thickness of the intima, a well-developed muscular media, and a thin adventitia. The intima contains a subendothelial layer of connective tissue, next to which is a prominent elastic-hyperplastic layer and after this, in turn, a musculoelastic layer. The media is composed of circularly disposed muscular fibers which are accompanied by elastic fibers. The elastic fibers are delicate in the inner half of the media and coarse in the outer half. No distinct elastica externa is present. The adventitia is composed of collagenous and elastic fibers and is not well developed. The coronary arteries are richly supplied with vasa vasorum (Gould 1953). In comparison, superficially distributed arteries have many elastic elements, deeper intramurally distributed arteries show crest-like intimal thickenings (Robb 1965).

Segmentation of the RCA. The course of the RCA may be divided into four segments (Forssmann 1979): the proximal segment (including the origin of the conus branch), the anterior segment (including the right anterior ventricular branches and the right marginal branch), the posterior segment (including the right posterior ventricular branches) and the posterior IV segment (including the PIA and its posterior septal branches).

Segmentation of the Heart According to the Branching Pattern of the RCA. Reig and Petit (2001) postulate six segments for the right ventricle. The three posterior segments (postero-basal, postero-mesial, postero-apical) are irrigated almost exclusively by branches from the RCA. The anterior segments (antero-basal, antero-mesial, antero-apical) enjoy a dual arterial supply, i.e., by the RCA and AIA, where the latter nourishes more than 20% of the right ventricular myocardium.

Segmentation of the Heart According to the Branching Pattern of the LCA. DiDio and Rodrigues (1983) and DiDio (1989) postulated a left hemi-heart and – given the relatively constant distribution pattern of the LCA – suggested the division of left ventricular myocardium into at least four coronary segments: the anterior interventricular segment, the lateral segment, the left marginal segment, and the posterior interventricular segment.

Reig et al. (1993a) determined 12 segments of the LV (Selvester's method): three septal segments (basal, mesial, and apical) and nine ventricular segments (superobasal, superomesial, superoapical, posterobasal, posteromesial, posteroapical, anterobasal, anteromesial, and anteroapical); the three septal segments exhibited a type of irrigation which was almost constant and of a dual type.

Division into Hemispheres and Quadrants. DiDio and Wakefield (1972) divided the heart into atrial (superior) and ventricular hemispheres. Seen from above and, for purposes of simplification, disregarding the actual position of the heart in the human body, the atrial hemisphere was subdivided into quadrants and each quadrant into thirds. The following terminology was used: two right quadrants (one anterior and one posterior), and two left quadrants (one anterior and one posterior). In each quadrant, the thirds were named medial, intermediate, and lateral. Each quadrant is related to a certain number of atrial branches deriving from the RCA and CA.

4.1.1
The Right Coronary Artery

The main stem of the right coronary artery (RCA) arises from the upper part of the right anterior surface of the aortic bulb slightly below the level of the right aortic cusp. In the right para-aortic region it is embedded in epicardial areolar tissue. Before it emerges, the RCA proceeds toward the right and somewhat anteriorly, and courses between the pulmonary artery, to its left, and the right atrium, to its right. Reaching the coronary sulcus, it descends within it almost vertically to the right cardiac margin, then follows a curved course around the base of the RA into the posterior part of the coronary sulcus, where it approaches the crux cordis region. In about 60% of our cases, the RCA reached the crux and terminated a little to the left of it. In 10% of our cases, the RCA terminated near the right cardiac border and in a further 10% between the latter and the crux; more often (as seen in 20% of our cases) it may reach the left border, replacing part of the CA (Williams et al. 1989) (see Sect. 4.3.2; (Fig. 3).

The main stem of the RCA ended at the crux cordis in 10% of the cases examined by James (1961) and von Lüdinghausen (1978). It bent at a right angle and – now designated PIA – proceeded in the posterior IV sulcus towards the apex cordis.

The RCA gives off several branches which depart almost perpendicularly; their existence is not invariable (according to Banchi 1904, Spalteholz 1924, and Baptista et al. 1988).

According to McAlpine (1975) and Williams et al. (1989) the following branches, beside the conus branch, are of clinical importance: right SA node branch, RSSA, right atrial branches, preventricular branch, anterior ventricular branches, right marginal branch, (right) AV node branch (artery), right posterior ventricular branches, PIA, and posterior septal branches.

4.1.2
The Conus Branch

In 40%–50% of the population, the right aortic sinus has two arterial ostia of different sizes (Schlesinger et al. 1949; Barry and Patten 1953, McAlpine 1975). The second ostium is usually less than 1 mm in diameter and gives rise to the conus branch (James 1965), whose diameter ranges from 0.5 to 1.9 mm. The exceptional role of the conus branch is underlined by synonyms such as "second" right coronary artery or "third" coronary artery (Schlesinger et al. 1949, James 1961; Paulin 1983). It supplies the conal myocardium and the layers of fatty tissue covering it (Hadziselimovic 1982; Williams et al. 1989).

Strongly Developed Conus Branch. If the conus branch is strongly developed it releases preventricular and ventricular branches which may nourish more than the conal or infundibular myocardium of the outflow tract. If the conus branch releases the right marginal branch, it may serve a greater part of the anterior wall of the right ventricle (Schlesinger et al. 1949; Hackensellner 1956; James 1961; Rathor et al. 1973; Vlodaver et al. 1976; Angelini 1989; Williams et al. 1989).

Two Right Conus Branches. The instance of two right conus branches has been described by Hadziselimovic (1982). In cases exhibiting obstructive coronary atherosclerosis the right conus branch may play a significant role as an intercoronary bypass vessel. It commonly anastomoses with a similar left coronary branch to form a widened collateral, the "Circle of Vieussens" (James 1961; Baroldi and Scomazzoni 1965; Williams et al. 1989) (see Sect. 4.8.5).

In selective angiography the conus branch may be injected directly and may be understood as a hypoplastic RCA (Vlodaver et al. 1976).

4.1.3 Further Ramification of the RCA

Atrial anastomotic branch, Ramus atrialis anastomoticus, Kugel's artery. Some authors describe the relatively invariable appearance of an anastomotic branch, the so-called Kugel's anastomotic branch (Kugel 1927; McAlpine 1975; Hadziselimovic 1982); this artery arises from the proximal CA or proximal RCA and courses from anterior to posterior at the base of the interatrial septum (see Sect. 4.8).

Ventricular Branches. The significance of the conus branch as an often independently arising branch has been described in detail in Sect. 4.1.2.

When the conus branch does not arise from the aorta, it appears as the first branch of the RCA and supplies the right ventricular infundibulum. A strong conus branch may release a strong vessel to provide a larger area with nutrition – a so-called preventricular branch – whose branches are also called *preventricular branches* (Barry and Patten 1953).

The existence of a long (right) marginal branch near the right margin of the heart is a relatively regular finding. This vessel usually arises 1 cm or more (in a few cases up to 5 cm) proximal to the right margin, and proceeds along it most of the way to the apex of the heart.

Other branches to the anterior ventricular wall, less consistent in occurrence, are named *anterior ventricular branches* (Williams et al. 1989).

In a few of our cases there was an early origin of the right marginal branch at a few millimeters' distance from the aortic ostium of the RCA. In one case (with a dominant left coronary artery), the marginal artery with early origin was stronger than the RCA and terminated almost at the right cardiac border.

The *AV node artery* usually arises from the RCA just proximal to the origin of the PIA. It proceeds upwards to penetrate the atrial septum and supply the AV node (see Sect. 4.5.3).

The ventricular branches of the RCA on the posterior surface of the right ventricle are much shorter than those on the anterior surface. If the RCA terminates as the PIA (as it usually does), these short branches arise perpendicularly and supply a portion of the posterior wall near the AV sulcus; the remaining myocardium of the posterior right ventricle is supplied by the PIA and branches from the right marginal branch. Near the apex there is some supply from the AIA (see Sect. 4.6).

4.1.4
The Posterior Interventricular Artery

At its beginning, the posterior interventricular artery (PIA) forms a sharp, inverted loop; thereafter it continues in the crux and proceeds in the posterior IV sulcus towards the apex. In most cases the PIA terminates halfway between the crux and the apex (Hackensellner 1954; Bader 1963; Gregg and Fisher 1963; Baroldi and Scomazzoni 1965; Allwork 1979; Anderson 1982; Angelini 1989; Williams et al. 1989).

There was a single PIA in about 70% of our cases. In 6% there was a double PIA accompanied by parallel right or left branches, which were small and sparse. When the PIA existed alone it gave off a few branches, in particular to the right, but also to the left ventricle. In 10% of cases examined by Levin and Baltaxe (1972), von Lüdinghausen (1978), and Williams et al. (1989) it had been replaced by a left coronary branch.

The PIA gives off numerous posterior septal (perforating) branches to supply the myocardium of the IV septum, and the adjacent ventricular walls, including the posterior (posteromedial) papillary muscle (Estes et al. 1966a,b). The posterior septal branches are consistently much smaller than the anterior septal branches that arise from the AIA.

Early Division. In rare cases the RCA divides directly at the right margin into two rather large diverging trunks, the upper trunk supplying the upper right ventricle and crux area, the lower supplying the lower right ventricle. Similar cases show a PIA originating not in the crux cordis region but "earlier" in the coronary sulcus near to the right cardiac margin (James 1961).

There is a peculiarity: in rare cases the PIA is entirely replaced by a dominant long AIA (Levin and Baltaxe 1972).

4.1.5
Segmentation of the RCA

For practical reasons the RCA was divided into two segments: the anterior and the intermediate (Vlodaver et al. 1976; Williams et al. 1989). The first or anterior segment extends from the origin of the RCA to the right margin of the heart; it releases the right anterior ventricular and the right marginal branches and anterior atrial branches.

The second or intermediate segment of the RCA lies between the right cardiac margin and the crux. It releases one to three small right posterior ventricular branches, supplying the diaphragmatic aspect of the right ventricle.

DiDio and Rodrigues (1983) and DiDio (1989) postulated a tripartite segmentation of an (ideal) right hemi-heart which, in accordance with the relatively consistent distribution pattern of the RCA, comprised: (1) the conus segment, (2) the marginal segment, and (3) the posterior interventricular segment.

Corresponding to these three ventricular segments are six auricular segments related to the RCA (anterior medial, anterior intermediate, anterior lateral, posterior lateral, posterior intermediate, and posterior medial) and six auricular segments related to the LCA (posterior medial, posterior intermediate, posterior lateral, anterior lateral, anterior intermediate, and anterior medial). A feature common to the auricular branches and their segments is the relative inconsistency in the frequency of their occurrence (DiDio 1989).

4.1.6 The Left Coronary Artery

The main stem of the left coronary artery (LCA) arises from the aortic bulb just a little below the level of the free edge of the left aortic cusp. Embedded in areolar fatty tissue between the left auricular appendage and the pulmonary trunk, it describes a gentle curve around the latter and proceeds directly forwards and then to the left. After a length of usually 1–2 cm from its beginning, it divides beneath the left auricular appendage in two or more primary branches (James 1965) (Fig. 3).

The stem of the LCA, which is 3–5 mm in internal diameter, is larger in caliber than the RCA and supplies a greater volume of myocardium, including most of the left ventricle, IV septum, posterior papillary muscle, and left atrium, except in cases of a right dominant coronary artery, where the RCA partly supplies a posterior region of the left ventricle and the posterior papillary muscle (Estes et al. 1966a,b; Flaherty et al. 2001). The main stem of the LCA releases a great variety of ventricular branches; only the CA gives off a few atrial branches.

With the exception of the AIA and CA, the other branches of the main stem of the LCA arise at an acute angle. With little variation, the AIA supplies an average of 41.5% of the entire ventricular myocardium; the two left branches – AIA and CA – together supply 63.8% of the ventricular myocardium (Kalbfleisch 1975; Knez 2000).

Bifurcation

In 75% of our cases the stem of the LCA was found to bifurcate into two branches of approximately equal strength, the AIA and CA; this was also observed in 40%–55% of the cases examined by Hadziselimovic (1982) (Figs. 4a–c, 5a).

Trifurcation

In about 25% of our cases (in 35%–60% of Caucasian heart specimens according to Gross 1921, Crainicianu 1922, Kalbfleisch and Hort 1977, and Hadziselimovic 1982; in 38% of European; and in 74% of Bantu heart specimens according to Pepler and Meyer 1960), the LCA trifurcates and gives rise to three branches: the AIA, CA, and the median (intermedian, anterolateral, or diagonal) branch (Figs. 5b,c, 6b).

The median branch originates in the angle formed by AIA and CA. It is the most prominent vessel of the group of left-sided lateral branches (Gross 1921). However, it varies in its length and diameter. In a few cases it exhibits an epimural course of 2–3 cm length and continues its course intramurally; in others it has a shorter or longer intramural course (see Sect. 4.4.1).

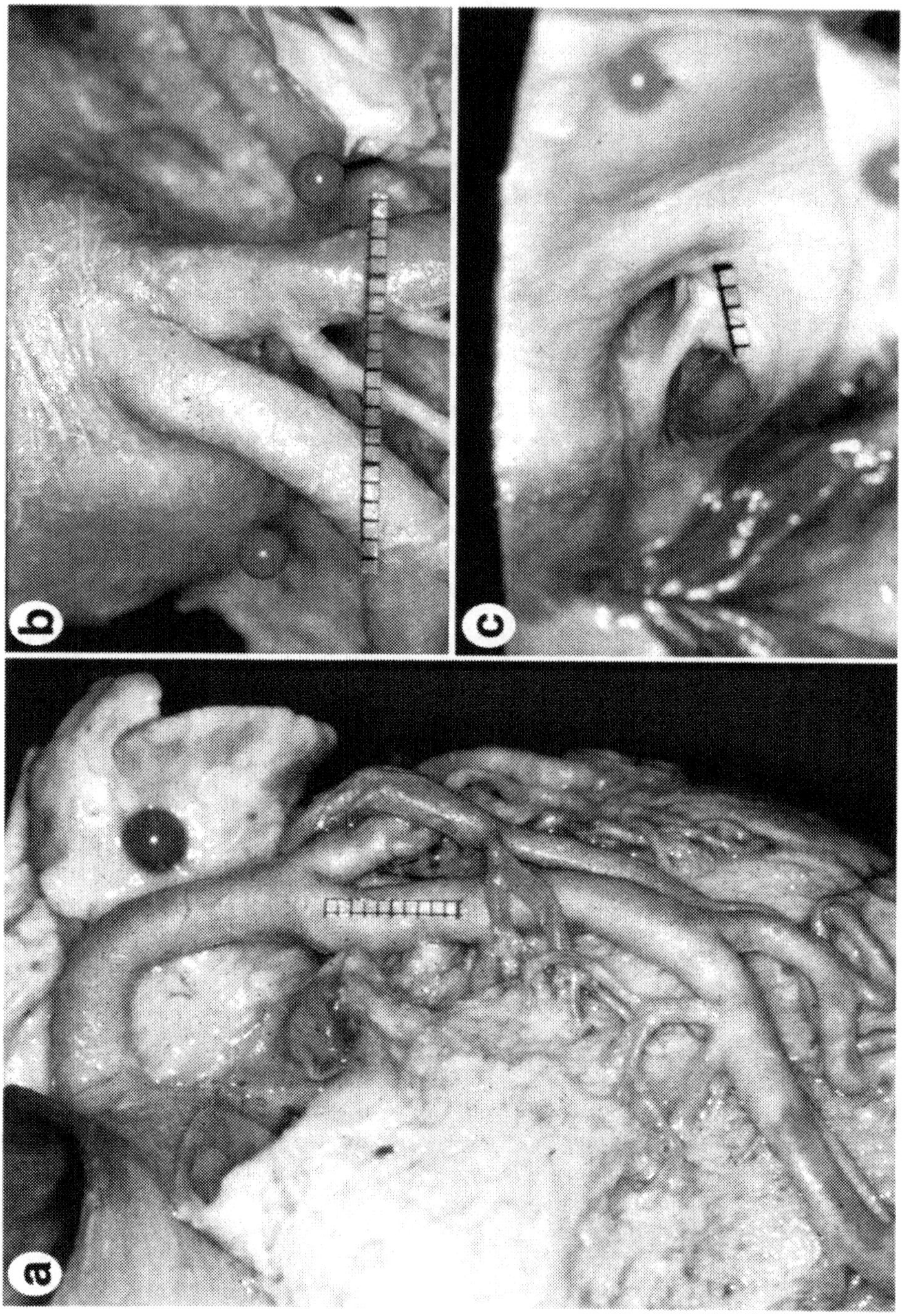

Fig. 4a–c.

Quatrufurcation

In very rare cases (4% of our specimens) there was a division of the main stem of the LCA into four branches of approximately equal size – the AIA, two median branches, and the CA – for the adequate supply of the anterior and lateral walls of the LV and the anterior IV septum (Fig. 6a).

Continuation of the Main Stem

In the majority of cases the stem continues its course directly as the AIA (in 74% of our cases), and less frequently as the median branch (20% of our cases) and rarely as the CA (in 6% of our cases, and "in rare cases" according to James 1961). At times the angle of division is even greater, so that the CA travels in a direction opposite to that of the parent artery (see Sect. 4.2.3) (Figs. 5a,b, 7a,b).

Ramification Pattern

The main branches of the LCA (i.e., the AIA and the CA) give off a great number of ventricular branches, although they may not always be extant (according to Banchi 1904; Spalteholz 1924, Baptista et al. 1988); left atrial branches derive only from the CA.

For practical reasons and due to their clinical significance, a few of the many ventricular and atrial branches are listed here, as they are in designated in English (according to McAlpine 1975; Williams et al. 1989).

Ventricular Branches

Anterior interventricular artery, anterior septal IV branches, anterolateral or diagonal branches, median branch, circumflex artery, (left) SA node branch, left marginal branch, posterior branch of the left ventricle. The infrequently present small left superior septal artery (LSSA) is either a first branch of the AIA or – in rare cases – a first branch of the CA.

Fig. 4a–c. Various lengths and modes of division of the main stem of the LCA as seen in cadaveric specimens. **a** An exceptional, long stem (26 mm) and acute angle of division into the CA and AIA (bifurcatio). The GCV crosses over the branches of the LCA. **b** An exceptional, short main stem of the LCA (designated "stemless") and acute angle of division into the CA and AIA (bifurcatio). **c** Interior aspect of the left aortic sinus (from the same specimen as in **b** exhibiting two arterial ostia. The ostium on the *left hand side* (lateral) belongs to the RCA, the ostium on the *right hand side* (anterior) to the AIA

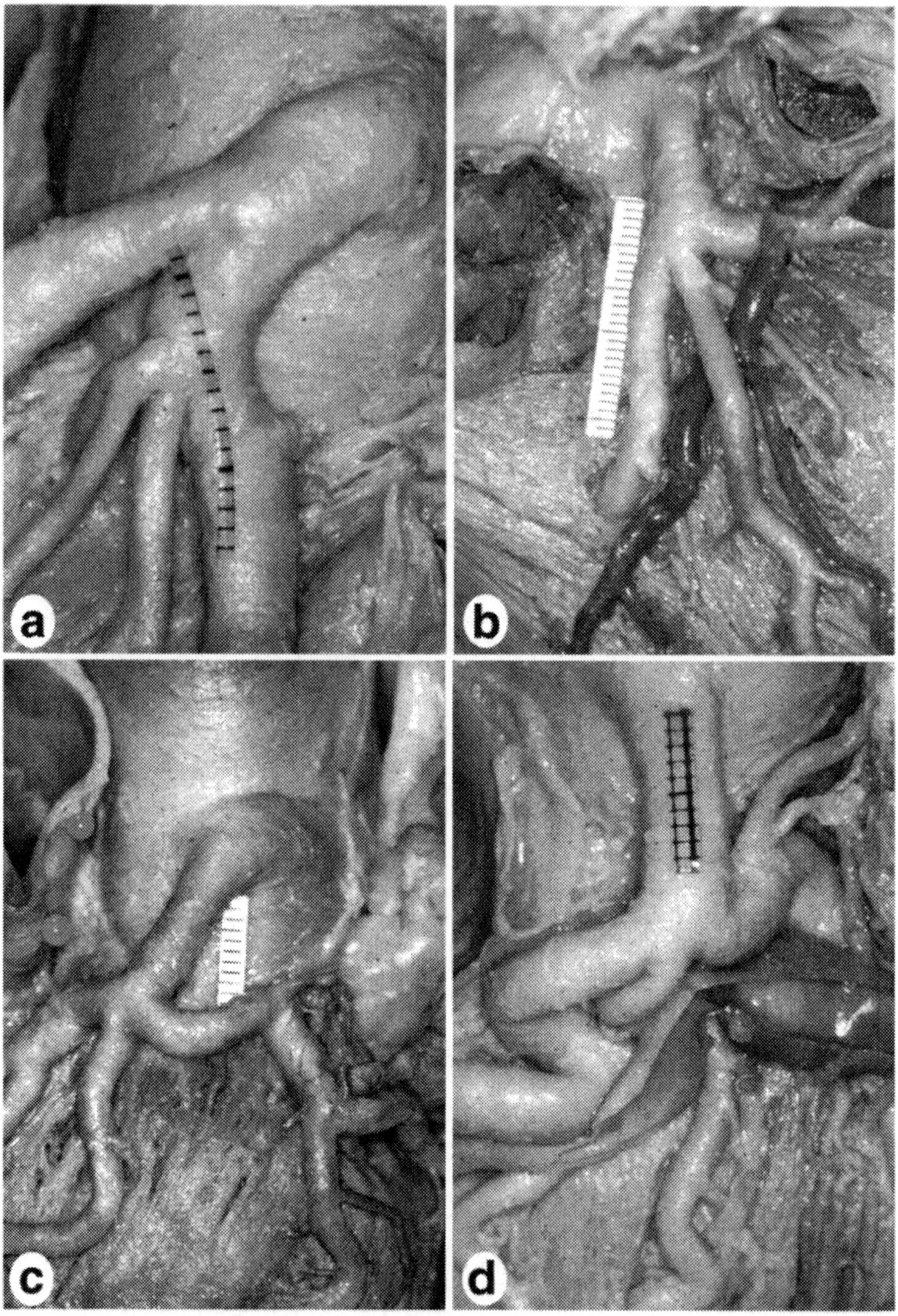

Fig. 5a–d. Angle of division of the stem of the LCA (lateral view). **a** Bifurcation and acute angle. **b** Trifurcation and right angle; the AIV crosses under the diagonal and over the circumflex branch. **c** Trifurcation and obtuse angle (almost 170°). **d** Trifurcation and obtuse angle (almost 190°); the AIV crosses over the diagonal and the circumflex branch

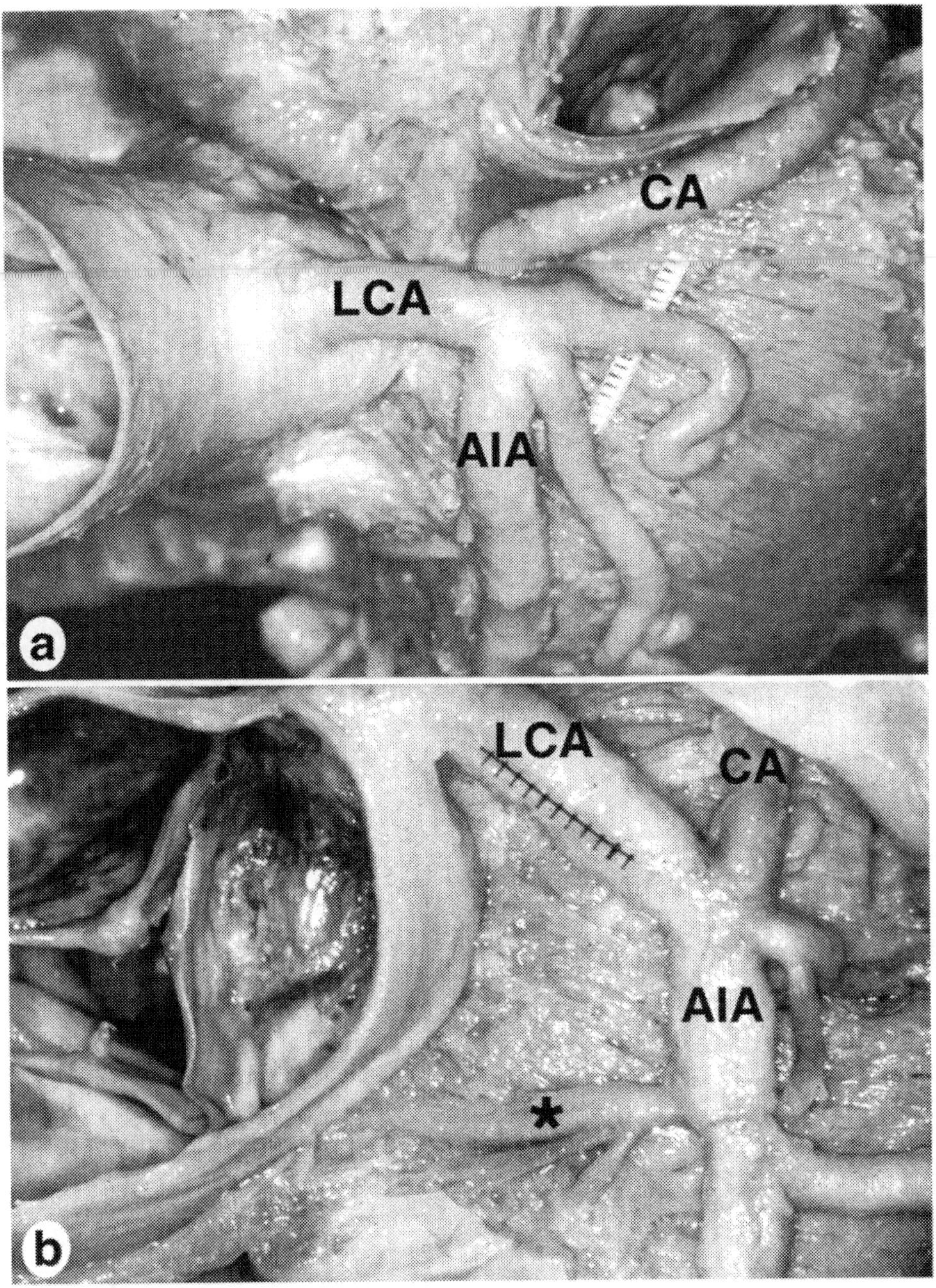

Fig. 6. a Quadrufurcation of the (long) stem of the *LCA* (superior aspect). **b** Cross-section of the aortic bulb (superior aspect) showing a rather posterior position of the ostium, and a rather long stem of the *LCA* with antero-lateral direction and an obtuse angle of division. A strong left superior septal artery – a branch of the proximal *AIA* – is marked by an *asterisk*

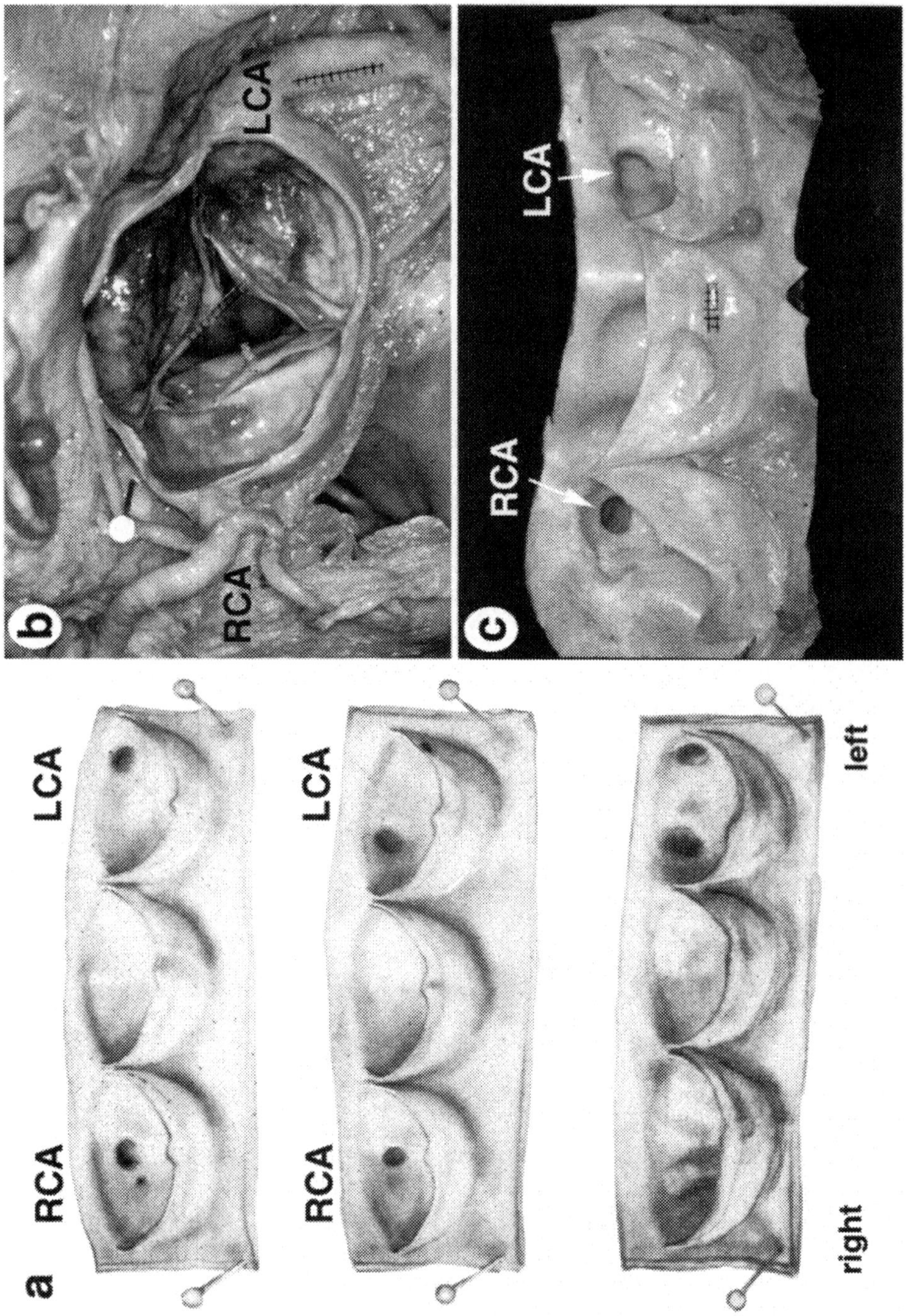

Fig. 7a–c.

Atrial Branches

(Left) sinuatrial node branch (variation), (left) lateral and posterior atrial branches, (left) AV node branch (variation).

4.1.7
The Anterior Interventricular Artery

The anterior interventricular artery (AIA) is a large and consistently found division (primary subdivision or segmentary artery). It is seen emerging on the anterior (sternocostal) surface of the heart between the root of the pulmonary trunk and the left auricular appendage (Gross 1921; McAlpine 1975; Angelini 1989). It is well developed in all hearts; it descends in a long and straight course within the anterior IV sulcus and in most cases reaches the apex. In many specimens, the AIA passes round the apex and ascends within the lower posterior IV sulcus for a distance of 10–50 mm (Gregg and Fisher 1963). The significance of the AIA for the supply of the anterior wall of the RV is of minor importance; however, up to 20% of the right ventricular myocardium is nourished by right ventricular branches of the AIA (according to Angelini 1989 and Reig and Petit 2001).

However, a (left) conus branch may occur near the origin of the first septal branch, proceeding towards the right ventricular infundibulum (Vlodaver et al. 1976). In rare cases an anteroseptal left ventricular infarction corresponds to a right ventricular infarction (Cabin et al. 1987). This means that in such cases the right ventricular branches of the AIA are larger than usual. During its course it may give off 1–3 lateral branches at acute angles to the left, towards the left margin of the LV, serving the sternocostal surface, and frequent anterior perforating or (intramural or deep) septal branches. The strongest branch supplying the myocardium of the right portion of the IV septum is an artery which arises from the AIA and descends intramurally and likewise perpendicularly in a straight course and reaches the moderator band and anterior papillary muscle. Because of its typical and consistent behavior it can be named a *left descending septal artery* (*LDSA*) or (according to Reig et al. 2000) *moderator band artery*. An obstruction of this artery may cause infarction of the anterior septum and of the right anterior papillary muscle combined with right bundle block and lethal septal perforation or rupture.

Fig. 7. a Position of the coronary ostia in the right and left aortic sinuses: (*above*) an accessory ostium of the conus branch in the right sinus; rather anterior position of the ostium of the *LCA*; (*middle*) a rather median position of the ostium of the *RCA* and posterior position of the ostium of the *LCA*; (*below*) a rare finding – coincidence of an empty right aortic sinus and two coronary ostia in the left aortic sinus (original drawing by J.S. Pupp, Würzburg, 1993). **b** Section of the aortic root in the horizontal plane (seen from above) showing the posterior position of both the *LCA* and the *RCA*, the antero-lateral direction of the stem of the *LCA* and postero-lateral direction of the *RCA*, and the respective acute angles of origin. **c** Posterior position of the ostia of both the *LCA* and *RCA*

4.1.8
The Median (Intermedian) Branch

This vessel is a variable branch and is defined by its distribution between the AIA and CA in a case of trifurcation of the LCA. In many cases it continues the direction of the main trunk and – after a short course – it dips into to the anterolateral wall of the LV exhibiting a rather long intramyocardial course.

4.1.9
The Circumflex Artery

The circumflex artery (CA) is the second main division (primary subdivision or segmentary artery) of the LCA and is comparable to the AIA in its caliber. The CA divides off from the stem of the LCA immediately at the point where the latter reaches the junction of the anterior IV sulcus with the (left) coronary sulcus. Both the origin and the initial few centimeters of the CA lie beneath the left atrial appendage, are almost embedded in fatty tissue and are not visible until the appendage is retracted. The artery emerges from beneath the appendage to a superficial position corresponding to the base of the mitral valve (Vlodaver et al. 1976) and proceeds within the coronary sulcus. The artery commonly extends to the left ventricular margin (margo obtusus), where it often terminates as the left marginal branch. Occasionally the CA passes round the margo sinister, to reach the diaphragmatic surface of the LV (Gross 1921; Angelini 1989; Williams et al. 1989).

The following, infrequently occurring ventricular branches (from proximal to distal), whose purpose is to nourish the lateral and posterior walls of the left ventricle and the anterior (anterolateral) papillary muscle, arise at acute angles (Estes et al. 1966 b; Flaherty et al. 2001): the median (intermedian) branch, the lateral diagonal branch, the left marginal branch, the posterior branch of the LV, and the (left) posterior IV branch; the latter is found only in a case exhibiting a dominant LCA.

The CA releases further small, unnamed branches which nourish the root of the aorta and atrial and ventricular myocardium adjacent to the coronary sulcus (so-called atrioventricular branches) (Barry and Patten 1953; DiDio 1967).

Atrial branches: the left SA node branch (at variable distance from the aortic ostium), "left atrial circumflex artery" (existence variable), posterior auricular branch, and posterior atrial branches.

In some specimens a large branch leaves the CA early in its course, rises to the base of the left atrium just above the coronary sulcus, and travels parallel to its parent counterpart. This has been named the "left atrial circumflex artery" (Paulin 1983). When extant this vessel often becomes the major CA (internal diameter more than 1 mm), the ventricular branch being diminutive. It is distributed along the lower portion of the left atrium and terminally over the posterior left atrial wall. Occasionally it releases the artery of the SA node. No vein was found accompanying the left atrial circumflex artery, although the oblique vein of the left atrium and the great cardiac vein are both located near it (as seen also by James 1961).

Some authors postulate a rather frequent appearance of an atrial anastomotic branch to a corresponding artery deriving from the RCA, forming the so-called Kugel's anastomosis situated at the posterior wall of the aortic bulb (Kugel 1927;

Silverman and Hurst 1968; Hadziselimovic 1982). In dissections, however, it was only found in 3% of our cases (see Sect. 4.8).

Functional Significance

The division of the LCA into the AIA and CA suggests an equality of supply which is false. The AIA alone supplies about half the muscle bulk of the left ventricle. In this sense, therefore, for a coronary circulation to be other than "left predominant" must be a rarity. The participation in the supply of the remainder of the left ventricle by the CA and branches of the RCA has, however, received rather more attention, because of the interesting variations in the pattern of coronary artery supply of the diaphragmatic (posterior) walls of the heart (Fulton 1965) (see Sect. 4.3).

4.2 Peculiarities of the Origin of Coronary Arteries

In terms of developmental anatomy (see Sect. 4.1), up to five coronary orifices may develop in the right aortic sinus; in rare cases the development of only one coronary orifice has been reported, either in the right or the left aortic sinus (May 1960; James 1961; McAlpine 1975).

Usually, there is a coronary ostium with a coronary artery rising from it situated within the corresponding aortic sinus, below or above the supravalvular crest. This ostial location allows maximal coronary filling during ventricular systole.

The ostia are situated above the reflections of the semilunar valves; in the human, the right coronary ostium lies 35° to the right and the left coronary ostium 65° to the left of the anteroposterior axis of the body (Gregg and Fisher 1963).

Three or four separate arteries with corresponding ostia can originate from the aorta without being considered anomalous (Angelini 1989). The use of the term "normal variant" is appropriate to describe the split origin of the conus branches or of the AIA and CA.

"Single coronary artery" refers to only one strong RCA or LCA originating from a single aortic ostium. An ectopic coronary origin is the result of the persistence of ectopic, additional, or atypical coronary buds, which connect with the RCA and LCA or one of their branches.

Coronary artery anomalies are found in 0.2%–1.2% of the population (Alexander and Griffith 1956; Bader 1963; Ogden 1970; Page et al. 1974; Engel et al. 1975; Kimbiris et al. 1978; Lipsett et al. 1994). The vessel which most commonly exhibits anomaly is the RCA (constituting 40% of all coronary arteries exhibiting anomalies).

4.2.1 The Location of the Right and Left Coronary Orifices in the Aortic Sinuses

The orifices are ordinarily the apertures in the wall of the aorta through which blood passes into the coronary arteries. These orifices are funnel-shaped depressions, some-

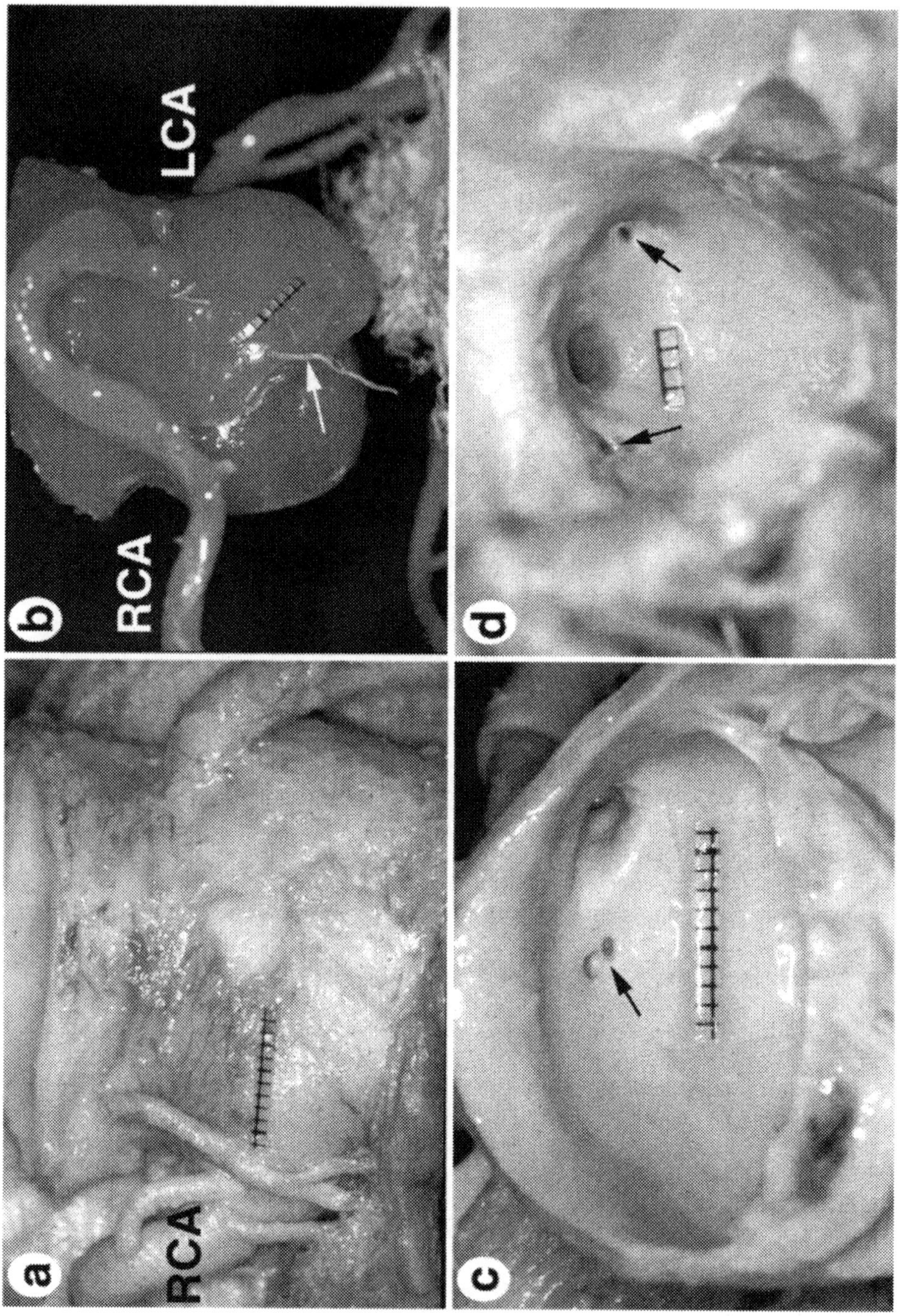

Fig. 8. a–d.

what rounded in contour, and ordinarily are found in the aortic sinuses, which are guarded by the right and left leaflets of the aortic valves.

In most atlases the ostia of the coronary arteries are shown as being situated exactly at the mid-point along each sinus close to the free edge of the aortic cusp (Rauber-Kopsch 1987; Dauber 1998; Sobotta 2000). However, there are variable positions for, and levels of, coronary artery origin (Figs. 6b, 8a–c).

In 56% of the cases examined by Vlodaver et al. (1976) and Paulin (1983), both coronary arteries arose below the sinotubular junction of the ascending aorta; in 30% the ostium of the RCA lay below the junction and that of the LCA above the junction; in 8% the ostium of the RCA lay above the junction and that of the LCA below, and in 6% both coronary ostia lay above the sinotubular junction.

Viewed in a horizontal plane through the junction line of the free edges and commissures of the cusps, one-third of the coronary orifices are located nearer to the posterior and anterior commissures of the corresponding aortic cusps rather than at the mid-point along the aortic sinus (Hadziselimovic 1982; Angelini 1989; Turner and Navaratnam 1996).

Viewed in an axial plane, the most common position is that both coronary ostia lie proximal to (below) the junction line (in 56% of the cases examined by Bankl 1977; in 30% the ostium of the RCA was slightly lower than that of the LCA lay above the junction line. In 8% the reverse was true, and in 6% both coronary ostia were above the line).

According to Williams et al. (1989) the coronary orifices are generally situated at the level of the cuspal margins; however, the right coronary orifice lay below it in about 10%, and the left coronary orifice in 15% of their cases.

The aortic origin of the main stems influences the orientation of the proximal segments of the LCA and RCA.

In cases he examined, McAlpine (1975) found that when the right aortic sinus was viewed in the horizontal plane, the ostium of the RCA was situated in a posterior position on its circumference in 44%, in an anterior position in 16% and in an intermediate position in 40%. Where the left aortic sinus was concerned, the oval ostium of the LCA was situated in the dorsal part in a posterior position on its circumference in 34%, in an intermediate position in 50%, and in an anterior position in 16%.

Seen in the sagittal plane, the coronary orifice was situated 1–2 mm below the supravalvular crest of the related cusp in 96% of the cases. Only in rare cases were

Fig. 8a–d. Multiple and accessory coronary branches arising from the area of origin of the *RCA* corresponding to multiple ostia in the right aortic sinus (**a**, **c**, and **d** are cadaveric specimens, **b** is a corrosion cast). **a** Anterior view of the aortic bulb showing the ascending stem of the *RCA* and the descending stem of the *LCA*. Four arteries are given off from the aorto-coronary junction: a right conal branch, an adventitial branch (vasa vasorum), and two right superior septal arteries (RSSAs). **b** Dextro-lateral surface of the aortic bulb showing the right and posterior aortic sinuses, the winding stem of the *RCA* and the descending stem of the *LCA*. A small RSSA (18 mm in length) arises from the lateral wall of the right aortic sinus and courses downwards (marked by an *arrow*). **c** Two accessory orifices at the midpoint along the right aortic sinus: one constitutes the origin of the conus branch, the other the origin of the RSSA. **d** Right aortic sinus showing accessory orifices in its anterior and posterior section which constitute the origins of the conus branch (*left arrow*) and the RSSA (*right arrow*). The ostium of the *RCA* can be seen at the midpoint along the sinus there variations in which it was situated in a position just above the supravalvular crest. These exhibited an anomalous high origin of the coronary arteries in the wall of the ascending aorta (McAlpine 1975) (see Sect. 4.2.7).

4.2.2 The Lengths and Diameters of the Major Coronary Arteries

In 92%–95.5% of autopsy cases examined by Angelini (1989), the LCA had a single initial stem or trunk of variable length (2.0–40.0 mm, mean: 13.5 mm) and size (2.0–5.5 mm, mean luminal diameter 4.0 mm) Figs. 4a–c, 5a–d, 6a,b).

Diverse estimates of the length of the main stem of the LCA have been made in recent clinical studies. *Anatomically*, it has been found to range from 1 to 26 mm (average 13.5 mm) before bifurcation into the AIA and CA branches (Baroldi and Scomazzoni 1965; McAlpine 1975; Penther et al. 1977; von Lüdinghausen and Valeske 1978); *angiographically*, it has been found to range between 7.5 and 20.5 mm (mean: 12.8 mm) (Lewis et al. 1970; Vlodaver et al. 1976; Waller 1983). The stem itself only gives off cardiac branches in very rare cases. Its direction varies with progressive age (see Sect. 4.2.3). A stem of the LCA of more than 20 mm length is designated an "unusual, long stem of the LCA" (Vlodaver et al. 1976) (Fig. 4a).

The LCA is considered "to be short" when it is less than 5 mm long. The designation "an unusual, short stem of the left coronary artery" often indicates that there are two orifices in the left aortic sinus (Fig. 4b,c) (see Sect. 4.2.4).

The AIA measures from 10 to 13 cm in length, whereas the usual nondominant CA measures about 6–8 cm in length. According to Waller (1983), a dominant RCA exhibits a length of about 12–14 cm before giving rise to the PIA.

The luminal diameters of the major coronary arteries at their origins in adults range as follows: main stem of the LCA, 1.5–5.5 mm (mean: 4.0 mm); AIA, 2.0–5.0 mm (mean: 3.2 mm); CA, 1.5–5.5 mm (mean: 3.0 mm); RCA, 1.5–5.5 mm (mean: 3.2 mm) (Baroldi and Scomazzoni 1965; Angelini 1989).

The magnitude of the diameter of the LCA exceeded that of the RCA in about 60% of the hearts they examined; that of the RCA was larger in 17%, and the size was approximately equal in 23%.

In postmortem angiographs performed by Köhler et al. (1981), the diameters of the coronary ostia averaged 4.83 mm (LCA) and 3.83 mm (RCA). There is a direct proportional relationship between hearts of weight up to 500 g and the sum of the diameters of their ostia (Köhler et al. 1981).

Measurements made on arterial casts or angiograms are of limited value, since the technique used is not always described, the physiological state of the heart specimen is often ignored, and no clear distinction is made between the measurement of external and internal diameters (according to Williams et al. 1989).

4.2.3
Three-Dimensional Orientation of the Coronary Stems, Angles of Origin of the LCA and RCA, and Angle of Division of the LCA

In adult hearts viewed three-dimensionally there are three angles of take-off of the RCA and LCA. They are measured in relation to the axis of the aortic root in three planes: seen from above (equivalent to the transversal plane), seen from lateral (equivalent to the sagittal plane) and seen from the front (equivalent to the frontal plane) (McAlpine 1975; Angelini 1989; Futami et al. 2001).

In the fetal heart, the stem of the LCA in the sagittal plane is directed antero-inferiorly and meets the surface of the aorta at an acute angle of almost 45° and the stem of the RCA describes an approximate right angle, coursing laterally, exactly to the right; in anatomical heart specimens from adults the stem of the RCA leaves the aortic root at an angle of 100°–110° thereby ascending slightly in a superio-lateral direction (102° according to Köhler et al. 1981).

Angles of Origin

The main trunk of the *LCA* takes off in the left posterior aspect of the aorta, at approximately 65° from the posterior axis, and courses anteriorly towards the anterior IV sulcus. Variations in location, however, are not rare and there is, perhaps, a standard deviation of 15° (Figs. 5a, 6a,b, 7b).

The *RCA* almost always arises from the right anterior surface of the aorta, is directed laterally and continues to the right coronary sulcus. May (1960) and Köhler et al. (1981) found the center-point of the angle of take-off to be approximately 35° from the anterior axis of the aortic root.

According to Köhler et al. (1981) from the variation in the angles of take-off in the vertical plane between the aortic wall and the proximal right and left coronary arteries it is evident that the LCA follows a descending course when compared with the RCA.

Reason for the Enlargement of the Angles of Origin

It is suggested that the reason for the enlargement of the angles of origin is the augmentation of the left ventricular myocardium in many individuals, and the increasing rigidity of the coronary arteries due to raised blood pressure in old age.

In cases where there is a high take-off of the artery from the coronary ostium the related angle of origin is always an acute one (Köhler et al. 1981).

A less frequent variant exhibits an almost tangential origin. In an unusual anomaly (Angelini 1989) of an intussusception of the coronary artery, the proximal arterial segment was embedded in the aortic wall.

The Angle of Division of the LCA

In children and young adults the angle of division is most commonly acute, whereas in the older individual it has become almost a right angle. In most subjects with pronounced hypertrophy of the left ventricle we found an obtuse angle of division (Figs. 5a–d, 6a,b).

4.2.4
Variable and Anomalous Coronary Ostia and Courses

With the development of coronary arteriography, it has been possible to recognize, and carry out the operative correction of, a wide range of variations and congenital anomalies. From among 7,000 patients who had undergone diagnostic coronary angiography, Kimbiris et al. (1978) and Cheitlin (1989) found an anomalous aortic origin of one or more coronary arteries in 0.64% (Figs. 8a–c, 9c,d).

During the course of research by the author, anomalies of the origin of the coronary arteries have been found more frequently in the dissection room, in 69% of male and 31% of female cadavers. An anomalous origin of the CA occurred in 58% of cases; in 27% both coronary arteries arose from the left aortic sinus, and in 9% both coronary arteries arose from the right aortic sinus.

The frequency of anomalies in almost 19,000 heart specimens examined by Alexander and Griffith (1956) amounted to around 2.8%; there were 39 cases of ostial anomalies and 15 cases exhibiting anomalies of course and distribution. Allwork (1987) found anomalies to be slightly more common in males.

Clinically unrecognized coronary variations and anomalies may lead to errors in diagnoses, and surgical problems may follow if an anomalous coronary artery is excluded from perfusion during open heart surgery, or if it is unwittingly incised by the surgeon.

Some reports indicate that certain anomalous coronary arteries are associated with anginal syndrome, myocardial infarction, and on occasion sudden death, even in the absence of occlusive arteriosclerotic disease (Benson 1970; Ogden 1970; Ogden and Goodyer 1970; Koops et al. 1973; Cheitlin et al. 1974, 1989; Kafrouni et al. 1981; Cheitlin 1989).

Ostial anomalies occur in the coronary arteries in three different groups of cases (Hadziselimovic 1982):

1. Where the number of actual coronary arteries extant is limited to one; this group includes those instances in which the origin of another coronary artery is from the trunk of the pulmonary arteries (Koops et al. 1973)
2. Where there may be up to five coronary arteries extant (Waller 1983)
3. Where there is an unusual origin of a number of coronary arteries in one aortic sinus and an abnormal distribution of some branches

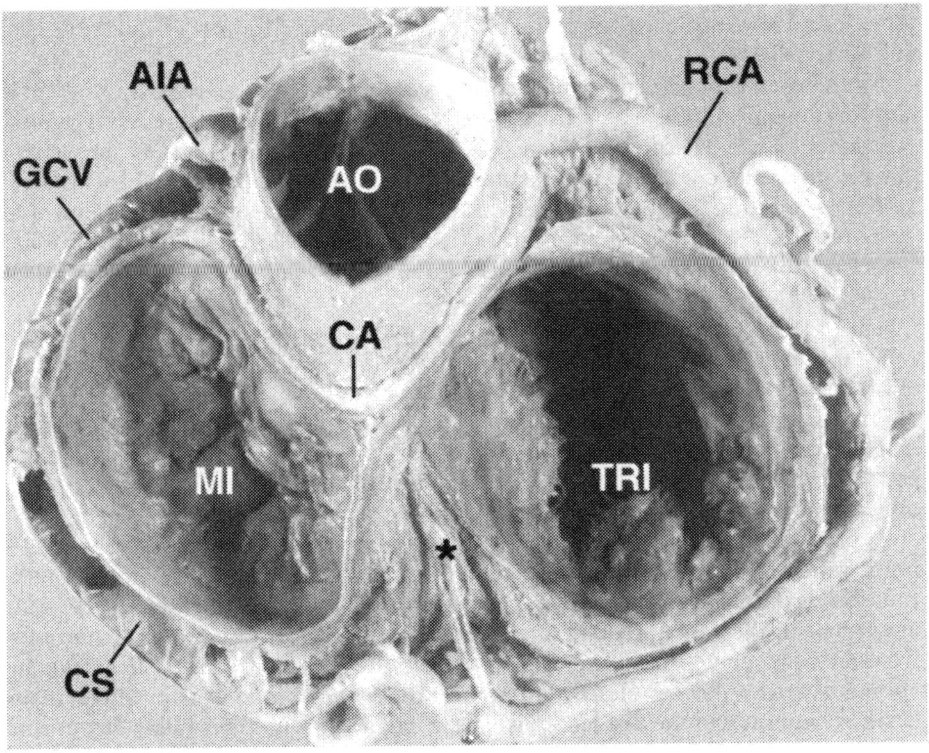

Fig. 9. Ectopic origin and aberrant course of the *CA* in a cadaver specimen seen from above (after removal of the atria)

4.2.5
Limited Number of Coronary Ostia: A Single Ostium for a Single Coronary Artery in the Left or Right Aortic Sinus

The incidence of a single arterial ostium in the aortic sinuses may be accompanied by one of two anatomical variations.

1. Myocardial perfusion is effected from the aorta via a single coronary artery.
2. There are two coronary arteries, one arising from the aorta and the other arising from the pulmonary trunk or, rarely, from one of its branches (Bookstein 1964; Doty et al. 1976; Vlodaver et al. 1975; 1976).

A Single Coronary Artery Providing Complete Myocardial Perfusion

A single coronary artery is found incidentally during postmortem examination. Therefore, it has been regarded as a minor coronary anomaly, which causes only slight disorder or has no clinical importance. Many individuals with this lesion are asymptomatic and have achieved longevity (Hallman et al. 1966; Vlodaver et al. 1975; Dollar and Roberts 1989).

Nowadays the instance of a single coronary artery has been revealed more often in large cardiological units which screen large numbers of patients. More recently, however, the anomaly has come to be considered of clinical relevance and to constitute a clinical phenomenon which is the potential cause of severe coronary heart disease such as myocardial ischemia, or even sudden death in younger patients (Allen et al. 1966; McAlpine 1975; Kafrouni et al. 1981; Click et al. 1988; Koizumi 2000). When detected, this anomaly is of prognostic importance (Sharbaigh and White 1974; Becker 1995).

The presence of a single coronary artery may thus be interpreted as signifying the total dominance of one artery over the other (Vlodaver et al. 1972, 1975; Kimbiris et al. 1978; Kafrouni et al. 1981; Becker 1995; Koizumi et al. 2000).

Sometimes a single coronary artery is associated with other cardiac defects. In complex cardiac malformations a single coronary artery may originate from the right carotid artery, right brachiocephalic trunk, or from the inferior aspect of the aortic arch (Bankl 1977).

The incidence of a single coronary artery in patients has been reported in 0.011% of the general population by Alexander and Griffith (1956), and 0.13% of the population which has undergone angiographic examination by Chaitman et al. (1976). In about half of the cases where a single coronary artery was present, the latter arose from the right aortic sinus, and in the rest, from the left aortic sinus. Usually, a single coronary artery divides shortly after the aortic origin into two or three major branches running to the cardiac walls. These major branches may follow the normal course of distribution, or may course in the area usually supplied by the "absent" coronary artery. In the latter situation the branches reach the area of distal distribution by crossing behind the aorta, between the great vessels and anterior to the pulmonary artery across the right ventricular outflow tract, to resume the normal epicardial distribution pattern (Ogden and Goodyer 1970).

A single "left" coronary artery with a single, large orifice in the left aortic sinus bifurcates into the AIA and CA in the normal way. In such a case the sternocostal walls are nourished by the branches from a strong conus branch traversing the upper part of the infundibulum to supply the upper anterior region of the right ventricle. The strong CA reaches the diaphragmatic surface and gives rise to the PIA in the posterior interventricular sulcus; thereafter it continues on its way, extending to and nourishing the lateral walls of the right ventricle (Koizumi et al. 2000).

A single "right" coronary artery is a large, dominant artery with trifurcation of its short stem.

1. It gives rise to a strong conus branch, to the regular RCA and the CA. The strong conus branch crosses the conus arteriosus or penetrates the infundibular part of the IV septum (in a so-called transseptal course) to become an AIA and supplies the entire sternocostal wall of the heart. The regular RCA supplies the right ventricular and atrial myocardium in the usual way. The CA courses retroaortically and epicardially through the pericardial sinus transversus to proceed within the left coronary sulcus (Dollar and Roberts 1989).
2. It gives rise to a "normal" RCA and strong, well-formed LCA. The latter subsequently courses between the aorta and pulmonary trunk or right ventricular infundibulum and finally divides into two vessels, the AIA and CA. The PIA is

well-formed, and extends out over and supplies the diaphragmatic (inferior wall) of the heart (Kafrouni et al. 1981; Liberthson 1989). This type of single "right" coronary artery is believed to be hemodynamically significant and to hamper myocardial perfusion (Levin et al. 1978).

Peculiarities

In rare cases the stem of a single coronary artery is lacking and either the AIA or the CA originate independently from the right aortic sinus (see Sect. 4.2.6).

In very rare cases a single coronary artery may have an ectopic origin from the pulmonary trunk. The prognosis here is fatal when this is the only source of myocardial nourishment (Hoganson et al. 1983).

Coincidence of a Single Coronary Artery with an Aortic Origin and a Second Coronary Artery Arising from the Pulmonary Trunk

The coincidence of a "single coronary artery" with an aortic origin and a second "single" coronary artery arising from the pulmonary trunk may be the reason for a left-right shunt of the blood flow and an arterio-venous fistula; it may be observed when one of the embryonic coronary arterial buds arising from the pulmonary artery has persisted.

The origin of a main coronary artery from the pulmonary trunk is a serious anomaly. The most commonly encountered form involves the LCA. The instance of the LCA arising from the pulmonary trunk and then assuming the course and distribution pattern of a coronary artery of normal origin is designated as Bland-White-Garland syndrome (Gouley 1950; Hackensellner 1955; Bookstein 1964; Jokl et al. 1966; Bankl 1977).

Coincidence of Normal Coronary Arteries and Congenital Arterial Communication with a Chamber of the "Right Heart"

An arterio-venous fistula between a "normal" coronary artery with an aortic origin and one of the right heart chambers (right atrium, coronary sinus, right ventricle) results when a communication between an intramural arterial branch and subendocardial sinusoid persists and enlarges (Edwards 1964; Wright 1970; Horiuchi et al. 1971; de Nef 1971; Bankl 1977) (see Sect. 4.1).

A left to right shunt may develop, with flow from the RCA through collaterals and retrograde flow in the LCA to the pulmonary artery – this constitutes a coronary artery "steal" syndrome which may result in the marked impairment of left ventricular perfusion and myocardial ischemia.

Clinically, the anomaly occurs in children in two forms (Bookstein 1964; Koops et al. 1973; Doty et al. 1976):

1. The severe infantile form with symptoms of cardiorespiratory distress and the threat of early death; surgical treatment is required.
2. The intermediate form with mild symptoms and growth retardation.

Adult patients may present symptoms at a late stage, or even never, but exhibit electrocardiographic evidence of myocardial infarction.

Peculiarities

The congenital, anomalous origin of only an AIA from the pulmonary trunk is a rare finding (Donaldson et al. 1979).

Of extreme rarity is the instance of the origin of a single coronary artery from sites other than aortic or pulmonary sinuses, for instance from the brachiocephalic trunk. Such a rarity is always associated with other severe cardiac malformations (Davies and Lie 1977). Small accessory coronary arteries may derive from one of the sinuses of the pulmonary trunk (Hackensellner 1955).

Associated cardiovascular anomalies (existence of a truncus arteriosus) and transposition of the great vessels (in two out of ten cases in which a "single coronary artery" was found) were described by Ogden and Goodyer (1970) and Vlodaver et al. (1976). The instance of a single coronary artery may be misdiagnosed as a complete ostial occlusion of one of the coronary arteries.

4.2.6 Greater Number of Ostia in the Left and Right Aortic Sinuses

Duplication of the ostia within *the left aortic sinus* is occasionally the consequence of a very short or even absent left main trunk and separated origin of the branches (AIA and CA). This phenomenon is also designated an "early" division of the LCA. It occurs in 1%–8% of otherwise normal hearts (Hackensellner 1956; Baroldi and Scomazzoni 1965; Vlodaver et al. 1976; Angelini 1989) (Fig. 4c).

In addition, there are a few cases of two ostia within the left aortic sinus when there is an anomalous origin of the RCA from it (Isner et al. 1984). There are also a few cases in which an RCA arises aberrantly from the left aortic sinus, passes between the aorta and the right ventricular infundibulum, and proceeds into the right coronary sulcus (Liberthson 1989).

A remarkable variety of arteries may develop in *the right aortic sinus*. In about 50% of humans an accessory artery, a "third coronary artery" (the conus branch), arises from a separate ostium in the right aortic sinus (Schlesinger et al. 1949; McAlpine 1975; Angelini 1989). Additional smaller ostia may be found there, which give rise to multiple ventricular and atrial branches. Up to five separate ostia have been described, for the (right) conus branch, the (right) SA node branch, the adventitial branch (vasa vasorum) of the aortic root, the preventricular branch, and the RSSA (Schlesinger et al. 1949; McAlpine 1975; Levin et al. 1978; Allwork 1979; Waller 1983; von Lüdinghausen and Ohmachi 2001) (Figs. 7a, 8c,d).

Among the latter, the conus branch is the most frequent "anomaly," because in almost 50% of cases generally, it arises independently from the aortic wall and its ostium is separate from that of the RCA. Waller (1983) described five relatively large orifices in the right aortic sinus: in addition to conus branches he found a duplicate AIA of the LCA. There are also a few cases of an anomalous origin of the main LCA from the right aortic sinus (Isner and Roberts 1978; Dollar and Roberts 1989; see next paragraph).

Origin of Both Coronary Arteries from One Aortic Sinus; Transposition of a Coronary Artery to the Other Aortic Sinus and Its Anomalous Courses

Most frequent in this group are specimens in which both the right and the left coronary orifice are situated together in either the right or left sinus of a (in many cases) bicuspid aortic valve (Keren et al. 1981). Such an anomaly is regarded as an incidental finding during autopsy; it is also considered to have minor clinical implications (Alexander and Griffith 1956; Edwards 1968; Benson 1970; Cheitlin et al. 1974).

When the *RCA* opens from *the left aortic sinus*, it turns to the right, courses between the pulmonary artery and the aorta and continues in the right coronary sulcus. The often slit-like opening of the anomalous RCA may cause decreased myocardial perfusion and clinically even infarction; the portion of the artery between the aorta and pulmonary artery may exhibit cyclic narrowing during systole (Isner et al. 1984; Ness and McManus 1988; Naidoo et al. 1988).

When the *LCA* opens from *the right aortic sinus*, the left main trunk (or the AIA or CA) may course posteriorly to the aortic bulb, between the aortic bulb and pulmonary conus (interarterial course), deep within the infundibular part of the IV septum (a so-called transseptal course), or anteriorly to the pulmonary conus (Benson and Lack 1968; Benson 1970; Chaitman et al. 1976; Murphy et al. 1978; Ishikawa and Brandt 1985, Dollar and Robers 1989; Nerantzis and Koutsaftis 1998).

Occasionally, such an LCA is hypoplastic and this may become clinically evident in a youthful individual; such a patient may suffer from myocardial ischemia. It may also be the cause of sudden death in children, young athletes or soldiers (Jokl et al. 1962; Cheitlin et al. 1974; Ishikawa and Brandt 1985).

Quite often one of these "transposed" coronary ostia is situated superior to the supravalvular crest, and exhibits a "high take-off" position (see Sect. 4.2.7). This may be the cause of sudden death.

Origin of the CA from the Right Aortic Sinus or from the RCA

The abnormal origin of the CA from the stem of the RCA occurs with a frequency of 0.1%–1.5% in subjects in the autopsy room or in patients who are undergoing coronary angiography (James 1961; Ogden 1970; Sanudo-Tejero et al. 1989; Karadede et al. 2001). From this origin the CA proceeds around the posterior aspect of the aorta and ultimately reaches the left coronary sulcus (Fig. 10).

In a classification of congenital coronary anomalies by Ogden (1970) this variation in origin is listed as minor.

The same applies when the CA originates from its own ostium in the right aortic sinus: such a variation is also relatively common (Wilson 1965; McAlpine 1975; All-

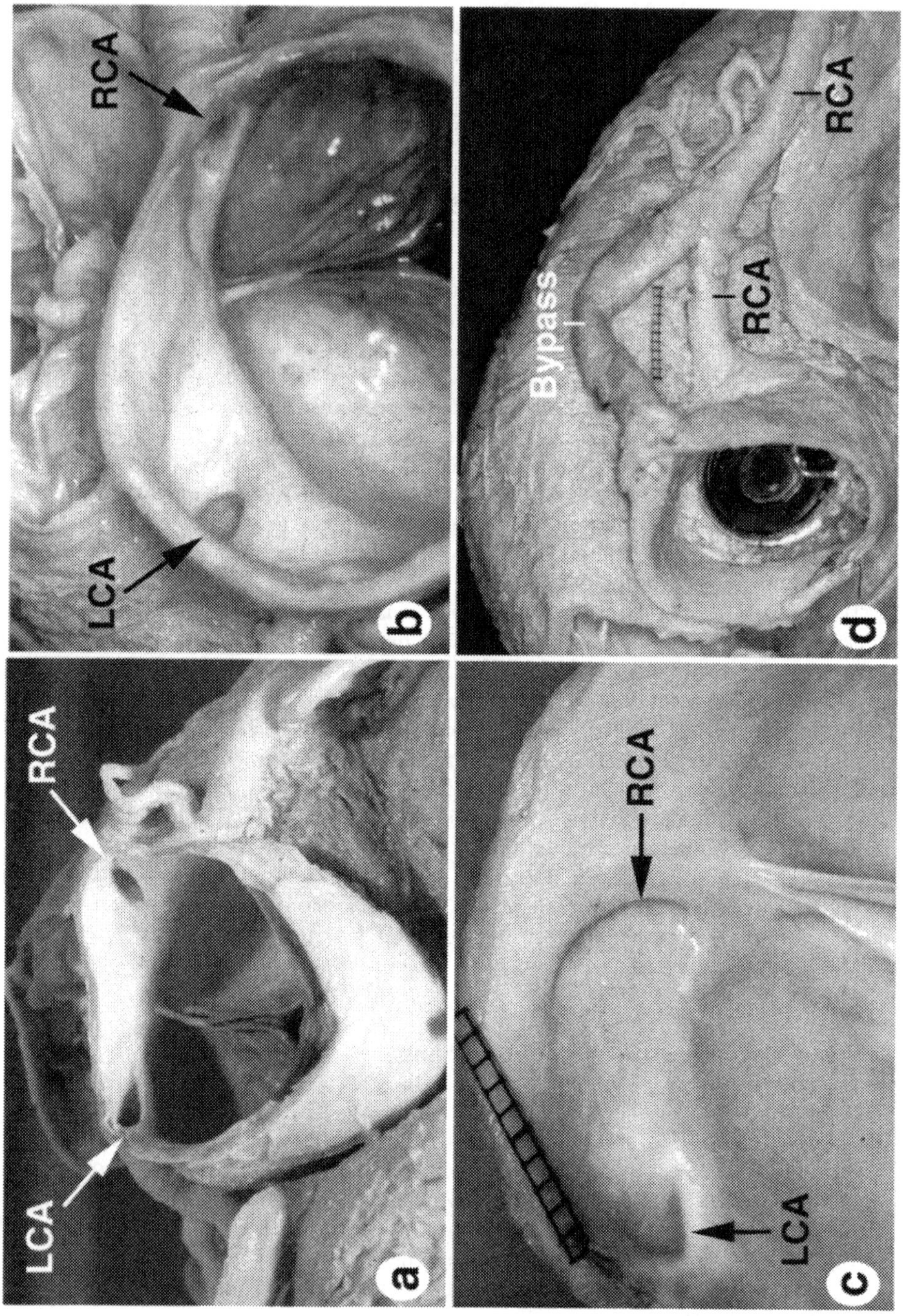

Fig. 10a–d.

work 1987; Click et al. 1989; Sanudo-Tejero et al. 1989) and best identified by means of aortography.

Complications – such as acute ischemia or infarction of the lateral wall of the left ventricle – may arise during an aortic valve replacement, when the RCA and LCA were cannulated during the operation and the cannula placed in the RCA has proceeded beyond the origin of the CA, and the abnormally coursing RCA is compressed by the valve prosthesis or inadvertently ligated by the circular suture (Vlodaver et al. 1976; Engel and Dalichau 1976).

Normal Coronary Arteries and Multiple Arterio-Ventricular (Arterio-Venous) Connections

While there may be a normal distribution pattern of the major coronary arteries in the subepicardial layers, multiple arterio-ventricular communications may occur when intramural arterioles are interconnected with remnants of embryonic sinusoids or widened arterio-luminal and arterio-sinusoidal channels (part of the group of smallest cardiac vessels).

An arterio-venous fistula is a rare but severe disorder; it has been reported in about 0.2% of instances of cardiac catheterization in adults (Edwards 1958; Coussement and de Geest 1994). In most cases the communication appears as a single fistula attached to the right atrium, ventricle, or the pulmonary trunk. Fistulae attached to the left ventricle or atrium, and especially multiple microfistulae involving both major coronary arteries, are very rare. Occasionally they cause myocardial ischemia due to intracoronary steal (Floyd et al. 1970; Eie and Hillestad 1971; Rose 1978; Liberthson et al. 1979; Duckworth et al. 1987; Coussement and de Geest 1994).

4.2.7 Aberrant or Ectopic (Superior) Origin, "High Take-Off" of a Coronary Artery

An origin of the RCA from the ascending aorta, usually 5 mm or more above the supravalvular crest, is a fairly common variant (Bader 1963; Vlodaver et al. 1975; Anderson and Becker 1982; Menke et al. 1985).

In 1% of all cases there is a slightly higher or ectopic location of a coronary orifice, superior or extrasinal (McAlpine 1975) to the supravalvular crest of the related aortic sinus. The RCA exhibits a higher frequency of the instance of ectopic ostium than the LCA (Hadziselimovic 1982) (Figs. 11a–c, 12a–c).

Fig. 10. **a–c** "High take-off" positions of coronary ostia in cadaver specimens. **a, b** "High take-off" position of the right and left coronary ostium above their ostia. **c** "High take-off" position of the right and left coronary ostium just above the left aortic sinus. Slit-like, oblique penetration of the proximal *RCA* through the aortic wall. **d** "Deep take-off" position of a right coronary ostium occluded during an operation for aortic valve replacement; an aortic-coronary bypass was implanted to prevent intra-operative myocardial infarction

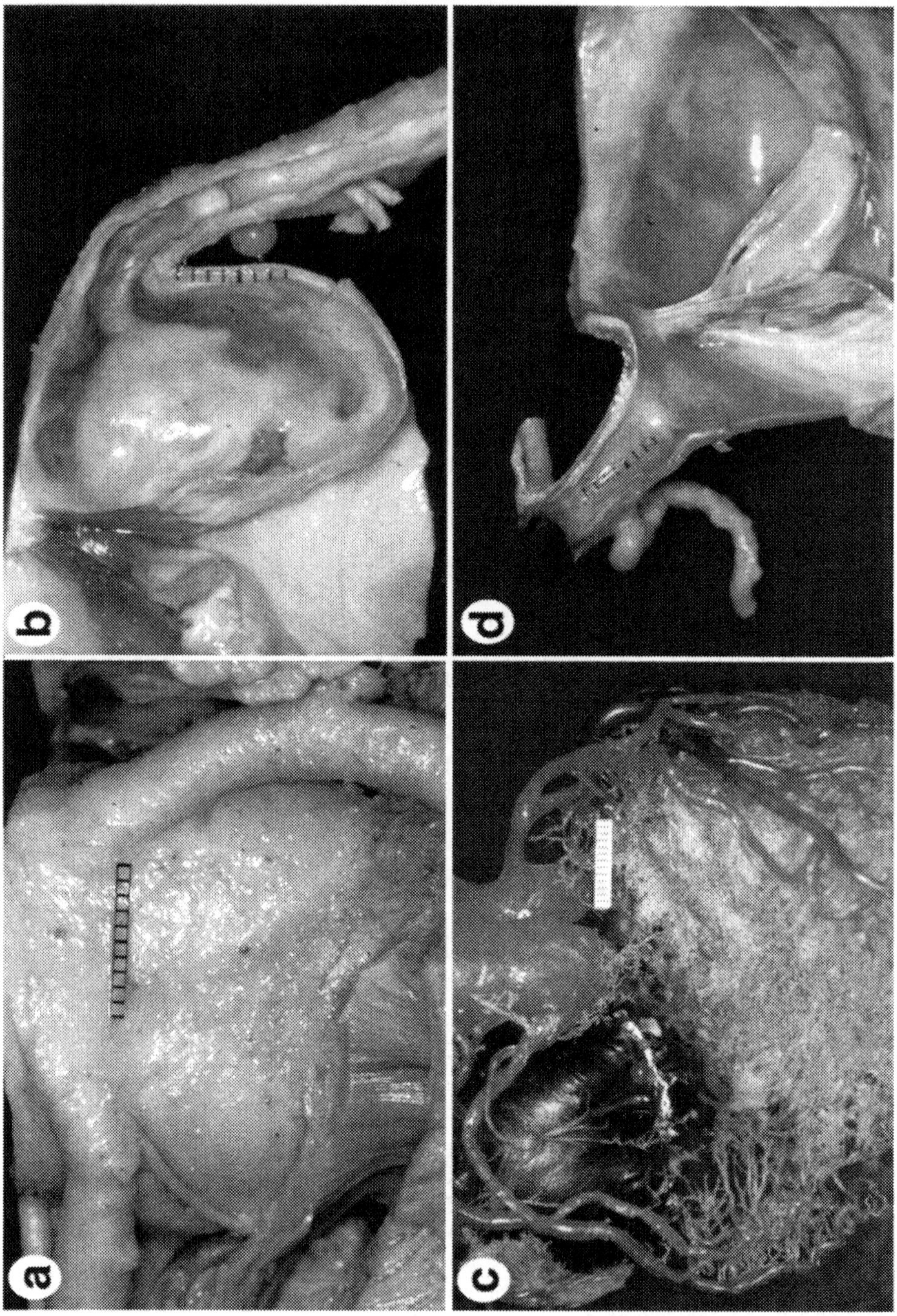

Fig. 11a–d.

A rather "high" origin and unusual course of both coronary arteries was recently described by Nerantzis and Marianou (2000). In this case the orifice of the RCA was found 19 mm and the orifice of the LCA 7 mm above the supravalvular crest of the left wall of the ascending aorta.

Piegger et al. (2001) reported an extremely high origin of the RCA 38 mm above the supravalvular crest. The artery followed a long intramural course within the aortic wall; at the site of the normal origin for the RCA, the anomalous vessel, still with a diameter of 2.5 mm, pierced the aortic wall and continued within the coronary sulcus.

In any high take-off position (i.e., 1–38 mm above the crista supravalvularis) of a coronary orifice, the first millimeters of the transmural artery may be narrowed or even occluded by a flap-like ostial margin which could be the cause of sudden death, even if the coronary arteries are devoid of atherosclerosis (Vlodaver et al. 1976; Mahowald et al. 1986). In its transmural section the artery passes obliquely through the aortic wall, thereafter to pursue a normal course (Allwork 1979, 1987) (Fig. 11b).

Cheitlin et al. (1974) pointed out that an aberrant LCA exhibiting a "high take-off" origin has an extremely acute angle of origin, which may compromise the arterial lumen.

Peculiarities

In rare cases there is a coincidence of a "high take-off" of a hypoplastic RCA, the normal course of a LCA, and also a hypoplastic CA (Menke et al. 1985).

Phylogenetic Aspects

A "high take-off" of one or both coronary arteries calls to mind the cardiac vascularization of the hearts of lower vertebrates. In species such as the amphibia and reptilia long coronary arteries arise "far" away in the cervical regions and course down to the walls of absorbed atrial chambers (Halpern and May 1958; Robb 1965).

4.2.8 Deep (Inferior) Origin, "Low Take-Off" of a Coronary Artery

In less than 1% of all cases there is a more inferiorly situated coronary orifice, deeply hidden in the right (in rare cases in the left) aortic sinus and immediately adjacent to the attachment of an aortic semilunar valve (Bader 1963; Angelini 1989).

Fig. 11a–d. Angles of origin of the RCA and LCA in relation to the axis of the aortic bulb. **a** Anterior view of the aortic bulb showing an acute angle of the LCA and right angle of the RCA. **b** Frontal section through the left aortic sinus showing an acute angle of origin of the stem of the LCA. **c** Anterior view of the aortic bulb (corrosion specimen) exhibiting a right angle of origin of the LCA and obtuse angle of the RCA. **d** Frontal section through the right aortic sinus showing an obtuse angle of origin of the RCA

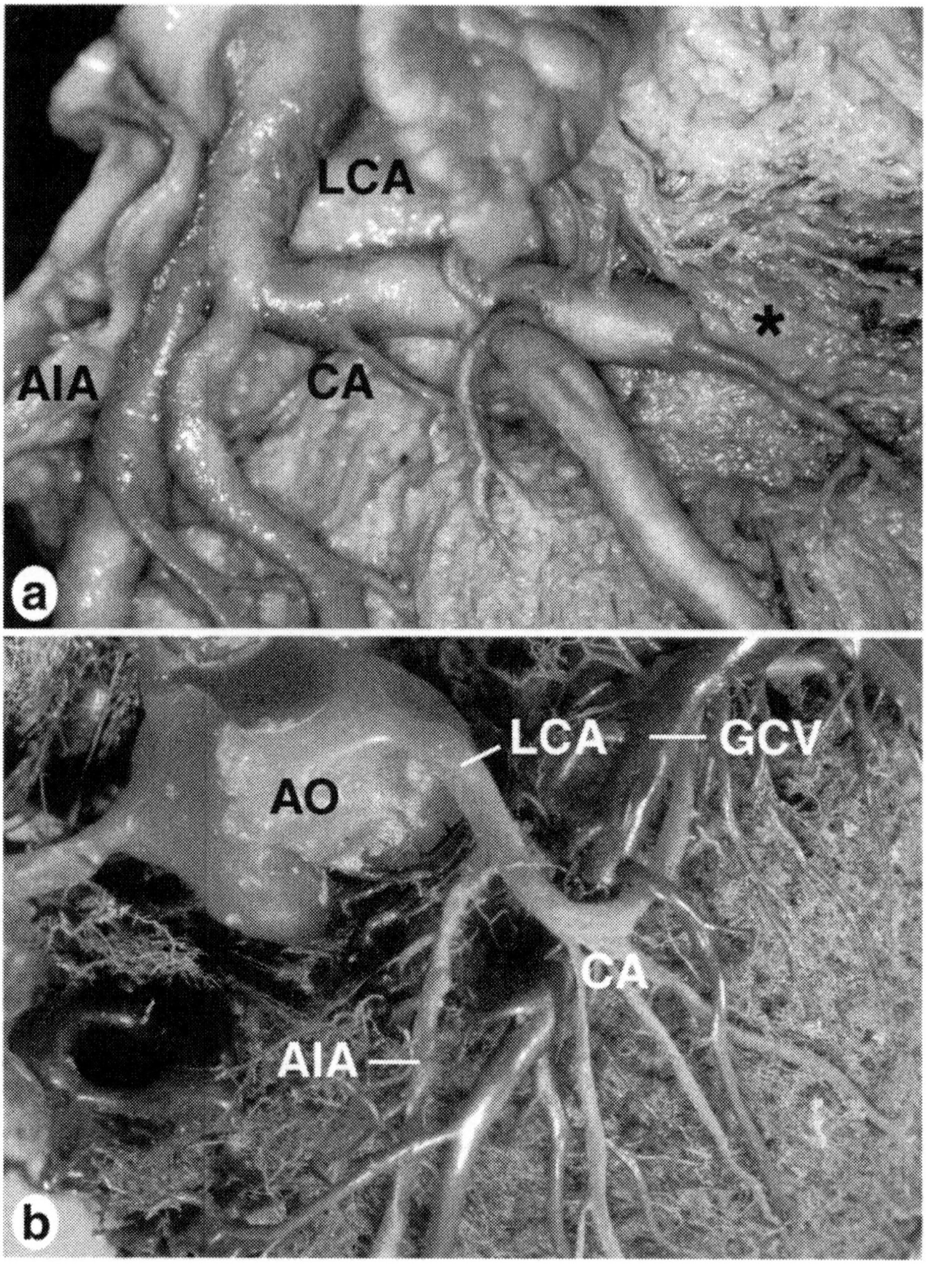

Fig. 12a, b. Continuation of the course of the stem of the *LCA*. **a** Anterior direction of the main stem and continuation as the *AIA*; right angle of division (cadaver specimen). **b** Lateral direction of a rather long main stem and continuation as the *CA*; acute angle of division; the *CA* releases three diagonal branches, the *AIA* none (corrosion cast)

During aortic valve replacement in a patient with the right coronary ostium at a deep location and a "low take-off" of the RCA there is a risk of iatrogenic ostial narrowing, occlusion by surgical sutures, and pressure of the transplant margin. In such a case myocardial ischemia and infarction will be registered during the operation and an aorto-coronary bypass will have to be implanted (Fig. 11d).

4.3
Coronary Dominance or Preponderance

The terms "dominant coronary artery," "coronary dominance," or "coronary preponderance" are commonly used and are related to the distribution pattern of the arteries between the right and left margins on the diaphragmatic surface of the heart (Schlesinger 1940a; Bargmann 1963; von Lüdinghausen 1975; Kalbfleisch and Hort 1977). According to Blunk and DiDio (1971), the mode of termination of the RCA and the CA and the relation of these arteries to the posterior IV sulcus are crucial to any decision regarding coronary dominance. The artery of dominance can only be determined on the latter evidence – which does not include a consideration of functional aspects – after examination of preparations in the dissection room (i.e., anatomically), or in the hospital (i.e., angiographically) (Schlesinger 1940; Blumgart et al. 1940; Blunk and DiDio 1971; Ahmed et al. 1972; von Lüdinghausen 1975; McAlpine 1975). The arterial supply of the left ventricular papillary muscles also depends upon the distribution pattern of the coronary arteries (Estes et al. 1966b). According to Reiss and Becker (1981), the level of the postero-medial papillary muscle is the "watershed" area between the right and left coronary arterial systems.

Definitions

The concept of coronary dominance or preponderance and its clinical significance was first proposed by Schlesinger (1938, 1940a,b). The dominant artery is the one which proceeds in the posterior IV sulcus and ramifies on the diaphragmatic surface of the LV (in a case of dominance of the RCA) or of the RV (in a case of dominance of the LCA), and also nourishes the posterior diaphragmatic portion of the IV septum (Figs. 13, 14a–d).

1. In a *balanced or normal coronary circulation*, the posterior wall of the RV and the posterior diaphragmatic portion of the IV septum are nourished by the RCA, and the diaphragmatic surface of the LV by the LCA and its branches alone.
2. A *dominant RCA (right coronary preponderance)* is a strong vessel which terminates as a PIA and gives off either a few or a number of branches serving a greater part of the posterior wall of the LV or almost the entire posterior wall.
3. A *dominant LCA (left coronary preponderance)* is present when the PIA derives from the CA. In such a case the LCA supplies the myocardium of the entire left ventricle and IV septum.

Some inconsistencies have arisen with regard to the percentages of occurrence of the various distribution patterns of the coronary arteries; only a small number of authors have maintained that in the case of a really balanced pattern, both right and left

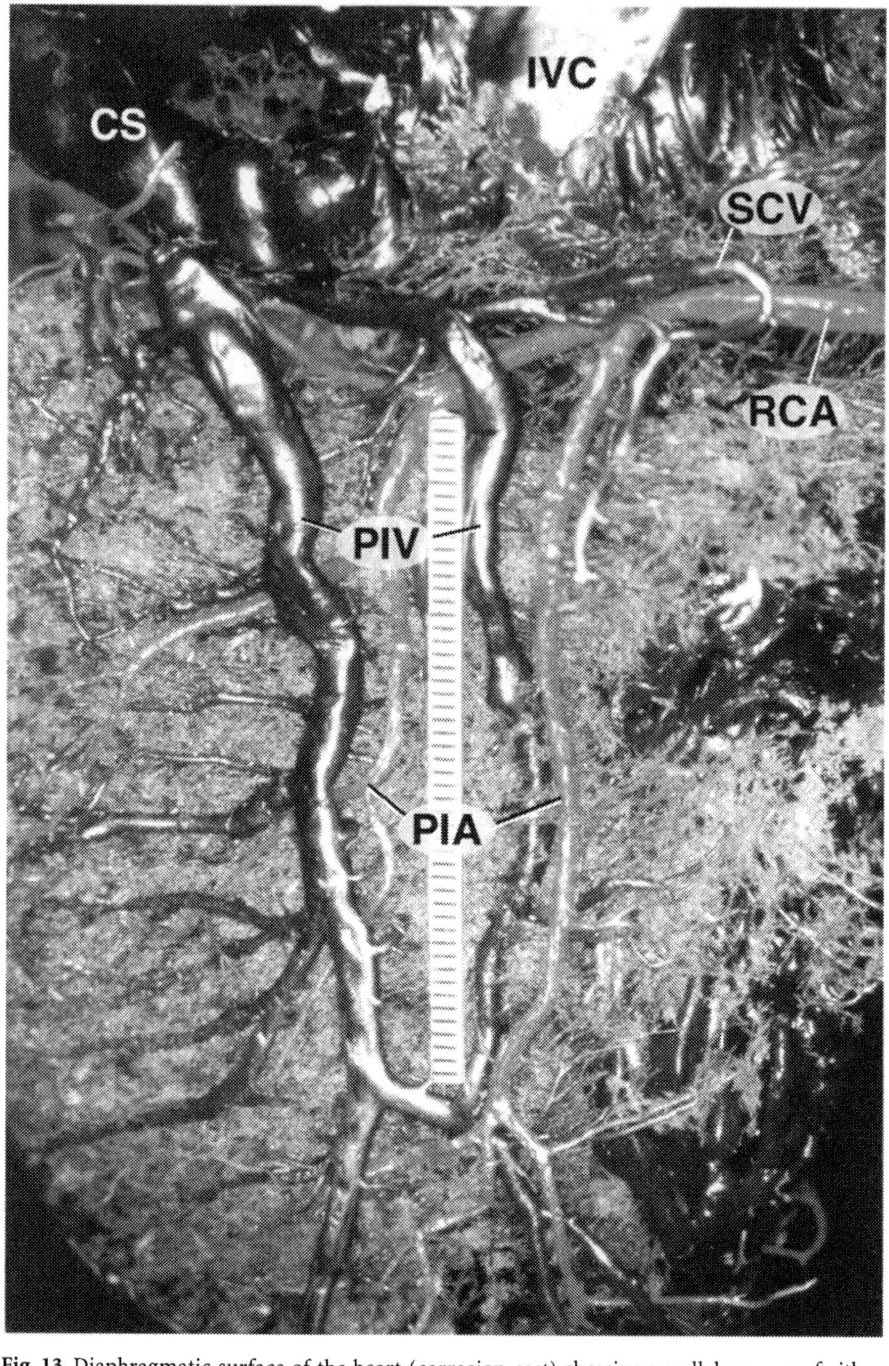

Fig. 13. Diaphragmatic surface of the heart (corrosion cast) showing parallel courses of either two *PIA*s and two *PIV*s. The double arrangement of the posterior interventricular vessels resembles the dual vascular supply of the right and left part of the IV septum. In the crux cordis area, the veins cross over the arteries

coronary arteries reach the crux area and both terminate in parallel-coursing interventricular arteries, i.e., in a right and left PIA. The septal branches of these two PIAs together supply the posterior part of the IV septum (Campbell 1929; McAlpine 1975; DiDio and Wakefield 1975; Kalbfleisch and Hort 1977; Harlan et al. 1983).

The size of the right and left coronary orifice is proportional to arterial predominance. Even in a case of a dominant LCA it is unusual for its branches to supply much of the posterior right ventricular wall, whereas – as mentioned before – it is usual for the branches of the RCA to supply a larger portion of the posterior left ventricular wall (James 1961).

However, there still exists controversy as to what exactly is meant by the arterial supply of a "large" or a "larger" portion of the posterior wall of the LV (Reiss and Becker 1981). This is the reason why the percentages of the occurrence of right or left dominant or balanced coronary arteries vary so much in the literature.

Differentiation of Coronary Dominance

Blunk and DiDio (1971) and Reiss and Becker (1981) suggest other criteria for the differentiation of coronary dominance which are concerned with the termination of the RCA and the termination of the CA in relation to the posterior IV sulcus. These authors described a right preponderance when the RCA proceeds in the posterior IV sulcus alone, as they found in 90% of their cases; in a few cases of right coronary preponderance branches of the RCA extended beyond the posterior IV sulcus and terminated in the diaphragmatic aspect of the heart to the left of the sulcus; there was a left preponderance when the CA originating from the LCA crossed the posterior IV sulcus and terminated beyond it, ending in the diaphragmatic aspect of the heart to the right of the sulcus, but did not supply significant parts of the right ventricular myocardium (Blunk and DiDio 1971).

4.3.1 Balanced or Intermediate Type of Coronary Distribution

A balanced or intermediate distribution pattern of the coronary arteries and myocardial supply was found in 55% of the cases studied by von Lüdinghausen (1978). In these cases the diaphragmatic surface of the heart exhibited one strong PIA or two PIAs, left and right, of equal size.

The right PIA was found by Campbell (1929) and Schlesinger (1938) to derive from the RCA in the crux cordis at a right angle and to proceed within the posterior IV sulcus, meandering in the direction of the apex cordis. The posterior septal branches supplied the posterior third of the IV septum; no branches of the RCA trespassed in the crux cordis.

The left PIA was found to arise from the RCA and course over the posterior wall of the left ventricle to supply this area and adjacent parts of the IV septum (Campbell 1929; Schlesinger 1938; Fulton 1965; Blunk and DiDio 1971).

The AIA of the LCA terminated adjacent to the apex cordis in 1/3 of the cases with a balanced coronary distribution studied by McAlpine (1975), in another third it reached the apex, and in the remaining third it coursed round the apex for one or two centimeters and ended in the apical posterior IV sulcus.

4.3.2
Dominant Right Coronary Artery

A dominant RCA (right preponderance) was found in 20% of cases studied by Williams et al. (1989) and 25% of the cases studied by von Lüdinghausen (1978) (Fig. 14a,b). The dominant RCA supplies not only the walls of the right ventricle but also a considerable part of the diaphragmatic wall of the left ventricle (Schlesinger 1940a); therefore, it extends with its branches beyond the crux cordis. It terminates by dividing into two or three main branches, namely the PIA, the posterior branch of the LV, and even the left postero-lateral branch. The latter runs under the posterior IV vein for varying distances and then proceeds over the lateral wall of the LV. Terminal branches of the PIA nourish the posterior papillary muscle and also, in a few cases, the posteromedial papillary muscle group (Estes et al. 1966b; Reiss and Becker 1981).

The left postero-lateral branch occasionally releases an accessory PIA, which may course over the diaphragmatic surface of the LV supplying its myocardium from its base to the apex (Vlodaver et al. 1976).

When there is a well-pronounced right dominance, the CA terminates as a left marginal branch which runs from the base towards the apex along the lateral wall of the left ventricle. When there is a moderately pronounced dominant RCA, the CA may be hypoplastic and terminate as a posterior branch of the LV or near the base of the crux (Vlodaver et al. 1976; Hadziselimovic 1982).

4.3.3
Dominant Left Coronary Artery

In 20% of our cases, there was a dominant LCA, i.e., a left preponderance, which falls into three groups (Fig. 14c,d) (Table 1):

1. In 9% of cases examined by von Lüdinghausen (1978) and 20% of cases examined by Williams et al. (1989) the CA of the LCA was quite a long, strong vessel, dividing at the crux cordis into the PIA and into an ascending branch towards the left atrium and the wall of the CS. It supplies the posterior wall of the left ventricle, the posterior third of the IV septum, and also the anterolateral and posteromedial papillary muscles (Estes et al. 1966b; Reiss and Becker 1981).
   There is an exception when branches of a dominant CA supply most of the myocardium of the posterior walls of the heart from the left to the right margin and the apical region (James 1961). In such cases the weak RCA has a short course, is rather hypoplastic and terminates near the right margin.
2. In a further 9% of our cases, the AIA was found to be a strong artery which gave rise to many anterior septal branches. In addition, after having coursed around the cardiac apex, the AIA gave off many inferior posterior interventricular septal branches. With these numerous anterior and posterior septal branches it supplies the entire IV septum and the anterolateral papillary muscle (Estes et al. 1966).
3. In 2% of our cases there were a few rare variations (see next paragraph, "Variations").

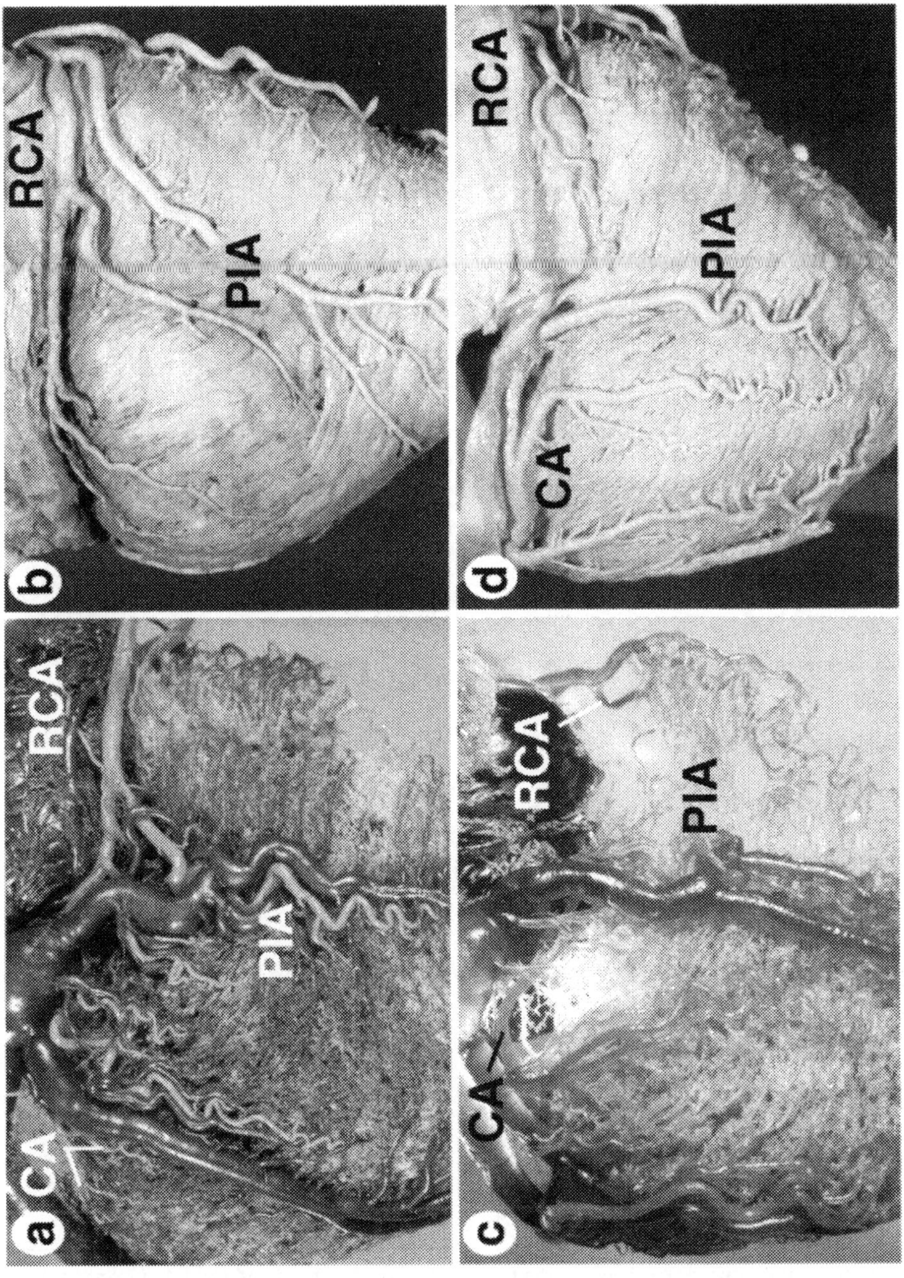

Fig. 14a–d. Various forms of dominant right and left coronary arteries as seen on the diaphragmatic surface of corrosion casts (**a** and **c**) and cadaver specimens (**b** and **d**). The branches of a dominant *RCA* are distributed on the posterior wall of the left ventricle. The *PIV* crosses over the arterial branches. The *CA* terminates as a left marginal branch. "Early" division of a dominant *RCA* at the level of the right cardiac margin into (1) the *PIA* and (2) the posterior branch of the left ventricle. There is a rather short intramural course of the *PIA* in the posterior IV groove. **c, d** A dominant *CA* releases a strong *PIA*, whereas the *RCA* is hypoplastic

Table 1. Frequency of distribution patterns of the coronary arteries in human hearts[a]

Author	Balanced type	Right dominance	Left dominance
Adachi (1928)	66%	26%	8%
Ahmed et al. (1972)	70%	40%	12%
n=94			
Blumgart et al. (1940)	40%	40%	20%
n=125			
Campbell (1929)	14%	72%	14%
n=100			
Crainicianu (1922)	75%	15%	10%
n=100			
DiDio and Wakefield (1975)	7.1%	73.5%	19.4%
n=98			
Fulton (1965)	30%	50%	20%
n=59			
Hadziselimovic (1982)	24%	63%	5%
n=100			
Kalbfleisch et al. (1974)	58%	29%	13%
n=100			
Lewis et al. (1970)	14.4%	69.5%	14.4%
McAlpine (1975)1	17%	71%	12%
n=100			
Paulsen and Vetner (1973)	71%	20%	9%
n=350			
Schlesinger (1940a)	34%	48%	18%
n=250			
Schoenmakers (1963)	68%	9%	23%
n=160			
von Lüdinghausen (1978)	55%	25%	20%
n=230			

[a]The conspicuous differences between the frequency of the balanced and right dominant type arise from the application of different definitions of the term "balanced type".

In these three groups with left preponderance the RCA was rather short and weak and often terminated as a short right marginal branch. Such a thin RCA only nourishes a small area.

Variations

In very rare cases (less than 1% of patients who have undergone angiography) there is an extremely dominant LCA (with a corresponding hypoplastic RCA) or an extremely dominant RCA (with a corresponding hypoplastic LCA) (Vlodaver et al. 1976).

Where there is an extremely dominant LCA its strong branches nourish all the walls of the left ventricle; where there is an extremely dominant RCA there is a mutual compensation: the branches of the right coronary artery supply the posterior walls of the left ventricle (Ogden and Goodyer 1970; Hadziselimovic 1982; Angelini 1989; Koizumi 2000).

Hypoplasia of the PIA may occur as an isolated entity associated with a large AIA which courses around the apex and ascends to the area usually supplied by the PIA (Vlodaver et al. 1976).

Phylogenetic Aspects

A study of the surface of the heart reveals that in most cases the RCA apparently dominates over the LCA. This can be explained as the manifestation of a persisting remnant from early phylogenesis (Halpern and May 1958).

In the dogfish, the sinus venarum is supplied by the coronary artery arising from the right hypobranchial artery. In the brook trout it is nourished by the dorsal coronary artery, and in the mud puppy by the ventral coronary artery. With the loss of gills and the incorporation of lungs, right coronary artery dominance becomes distinctly more apparent. The artery that nourishes the heart in the frog arises as a single vessel from the right carotid subdivision of the truncus arteriosus. This vessel migrates caudally and to the right and supplies the walls of the sinus venarum. In the wood turtle, right coronary artery dominance is also observed as the dorsal coronary artery migrates to the right and reaches and supplies the walls of the sinus venarum (Halpern and May 1958).

Significance and Functional Aspects of the Dominant RCA and the Dominant LCA

During intrauterine life there is both a right ventricular dominance and a dominant RCA (Gross and Kugel 1934). During childhood and adolescence, however, the LV has acquired thick myocardial walls. Therefore, any consideration of functional aspects has to take into account the fact that, in a case of preponderance of the RCA, the branches of the LCA still supply at least 50% of the ventricular myocardium; in a case of balanced coronary supply the LCA nourishes 60%–70% of the ventricular myocardium, while in a case of left coronary preponderance, that artery nourishes 80%–85% of it (McAlpine 1975).

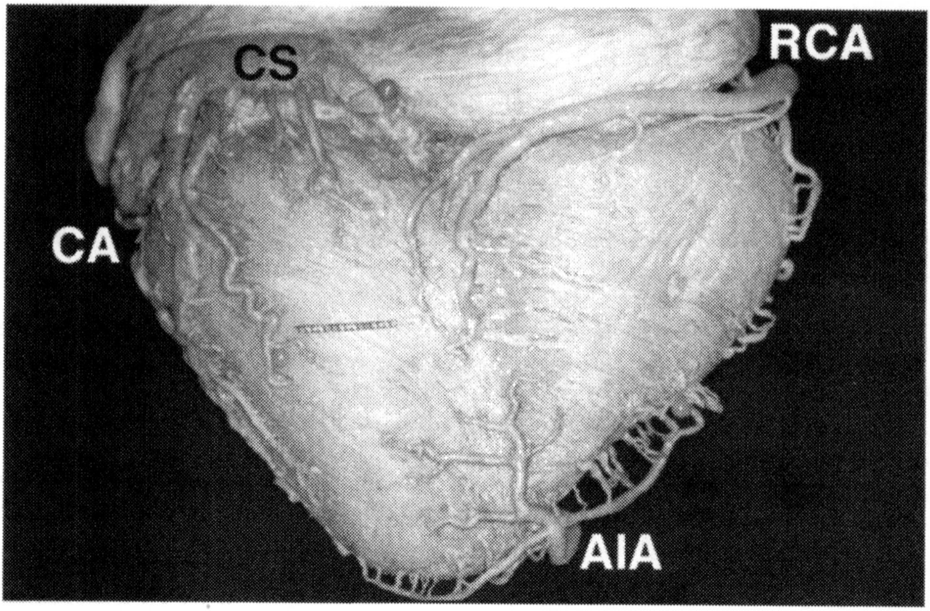

Fig. 15. Diaphragmatic aspect of a cadaver specimen showing a dominant *AIA* which courses round the apex region and supplies the apical parts of the right and left ventricles and of the IV septum

Depending on the individual ramification pattern of the coronary arteries on the diaphragmatic surface of the heart previously described a posterior left ventricular infarction, a posteroinferior left ventricular infarction and an apical and septal infarction (with participation of the corresponding papillary muscles) may be caused either by obstruction of the PIA of the RCA, PIA of the LCA, or AIA of the LCA (Fig. 15).

Numerous authors have compared the lengths and diameters of the right and left coronary arteries to establish which vessel apparently has more functional importance. However, the results are uncertain, inconsistent and contradictory.

4.4 Intramural (Intramyocardial) and Intraluminal Courses of the Coronary Arteries

In human hearts the main coronary arteries and their branches are presumed to course extra- or epimurally (subepicardially); the intramural or intramyocardial portion of the coronary arterial supply consists of numerous small branches that penetrate the wall in a perpendicular fashion and plunge into the myocardium only once, at their distal end. Frequently, however, one of the arteries exhibits an abnormal course and becomes intussuscepted into the myocardium (Angelini 1989) (Fig. 16a–d).

"Intramural courses" of coronary arteries or "myocardial bridges" of mural or tunneled arteries are the terms used to denote parts of coronary arteries which are quite often covered by strands of myocardium (Geiringer 1951; Vlodaver 1976; Williams et al. 1989). In other words, there are intramural or intramyocardial portions of the greater coronary arteries (Polacek 1961; McAlpine 1975; von Lüdinghausen 1975; Allwork 1987). Myocardial bridges, covering segments or parts of major coronary arteries and branches, have been found in 54%–80% of the cadaveric heart specimens investigated by von Lüdinghausen (1975), Bezerra et al. (1987), and Baptista and DiDio (1992).

The length of the intramural course (which equals the width of the myocardial bridge) ranges from 9 to 69 mm but is usually 10–20 mm; the intramural course may have a caliber of up to 5 mm (Polacek 1961). In rare cases a deep arterial course was found within the myocardium or even in the subendocardial layer of a heart chamber (von Lüdinghausen 1975).

4.4.1 Intramural Courses of Ventricular Branches

It is the AIA which most frequently follows an intramural course (in 55% of the cases studied). Such a course may be short (3–10 mm in length), medium-length (11–30 mm in length) or long (more than 30 mm in length) and usually begins shortly after the division of the stem of the LCA (von Lüdinghausen 1975; Bezerra et al. 1987) (Fig 16a).

The diagonal (antero-lateral) and intermediate branches of the LCA exhibited intramural courses in 33% of all the cases studied by Bezarra et al. (1989), the branches of the RCA in 22%, the CA in 15% and the main stem of the RCA in 1%–8%.

The arterial segment which is covered by a myocardial bridge may be divided into a prepontine, a subpontine, and a postpontine portion (Bezarra et al. 1989).

A short intramural course of an arterial segment makes a U-turn when it plunges into the myocardium for a few millimeters, turns back and comes up again to continue the otherwise straight epimural course.

Two or even three short courses in one specimen have frequently been observed.

Rarely, (in 3% of our cases) the AIA exhibits an extremely long course of up to 8 cm. In such a case the AIA plunges into the myocardium at the beginning of the anterior IV sulcus and does not come up again (von Lüdinghausen 1975). In one case reported by McAlpine (1975), an AIA of this type reached the subendocardial layers of the right ventricle transmurally (McAlpine 1975).

4.4.2 Long Intramural Courses

If an AIA or PIA follows an extremely long intramural course, the anterior and posterior IV sulcus will appear empty. If the main stem of the RCA or CA follows an intramural course, either the right or left coronary sulcus will be devoid of any artery (Fig. 16a).

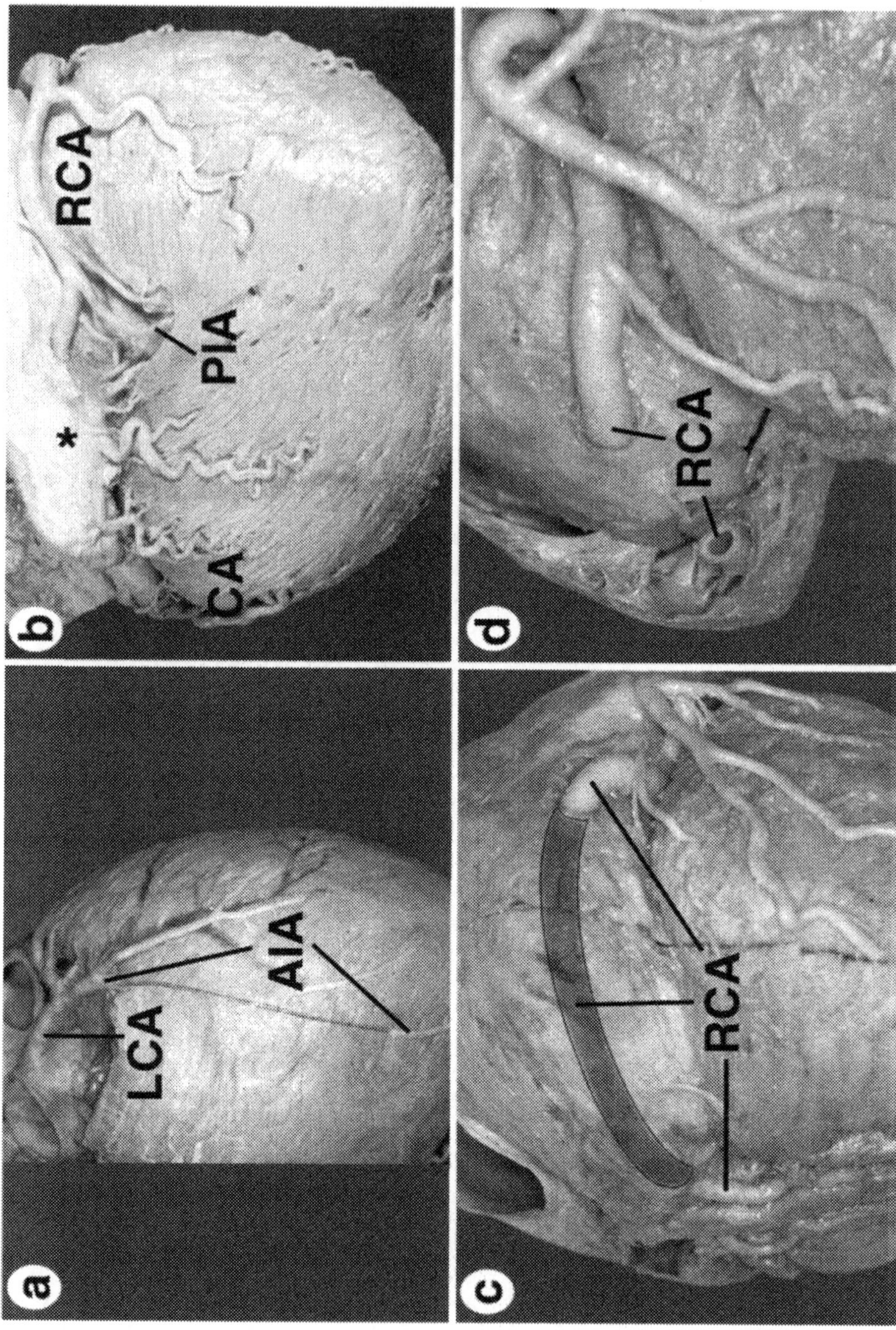

Fig. 16a–d.

Many right and left atrial branches follow intramural courses of up to 22 mm in length. The longest and strongest among them is the branch which supplies the SA node; it was a branch of the RCA in 65%, or of the LCA in 32%, of the cases examined by von Lüdinghausen (1978).

In one case the CA approached the CS and was – for a length of 4 cm – embraced and covered by the left atrial myocardium.

4.4.3
Subendocardial, Intraluminal, or Intracavitary Courses

Occasionally, a deeply coursing intramural AIA will reach the innermost layers of the right ventricle; this course is designated as subendocardial.

Occasionally, the RCA exhibits so-called intraluminal or intracavitary courses in the walls of the RA or LA (Fig. 16c–d).

Unexpected intramural courses of the coronary arteries constitute a high risk where any surgical procedure on the cardiac vessels is concerned.

4.5
Atrial Branches and the Arterial Supply of the Conduction System

During their course through the coronary sulcus, the RCA and the CA emit on average 3.5 small irregular branches to the anterior, lateral, and posterior walls of the right and left atrium (May 1960). Among them are the arteries which regularly supply the centers of the conduction system and one, two, or three atrial divisions of the so-called atrioventricular arteries (DiDio 1967; Vlodaver 1976; Williams et al. 1989).

The sinus venosi of the right and left atria (the so-called smooth portions of the atria) reveal an irregular pattern of small arterioles. The right and left auricles (the so-called rough part of the atria) exhibit parallel arterioles in the musculi pectinati, which supply clusters of capillaries. The wall of the left atrium contains arterioles which anastomose with each other and form a circular pattern around the orifices of the pulmonary veins.

Numerous arterial vasa vasorum are found in the terminal portions of the adventitia of the superior and inferior vena cava, originating from the arteries in the wall of the right atrium (Clarke 1965).

◄

Fig. 16a–d. Ventricular and atrial, intramural and intracavitary, courses of the coronary arteries. **a** Sternocostal surface of the heart showing a long, deep intramural course of a dominant *AIA* (marked by a millimeter scale) which can be assumed to run almost into the subendocardial layer of the anterior wall of the right ventricle. **b** Posterior surface of the heart with a short intramural course of the terminal section of the dominant *RCA* (marked by an *asterisk*) in the posterior wall of the left atrium. The posterior IV sulcus does not contain the *PIA* and may thus be designated artery-free. **c** Lateral view of the right heart showing a long intramyocardial course of the *RCA* through the meshwork of musculi pectinati of the right auricle (*marked*). **d** Incision of the anterior wall of the right auricle (of the specimen shown in c) to demonstrate the "intracavitary" course of the RCA

Atrial Branches and Atrial Quadrants

According to their subdivision of the atrial hemisphere into quadrants, DiDio and Wakefield (1972) found the total number of large, medium-sized and very tenuous atrial arteries (including the arteries for the conduction system) to be:

1. 79 atrial arteries in the right anterior quadrant (21.1%)
2. 46 atrial arteries in the right posterior quadrant (12.3%)
3. 137 atrial arteries in the left posterior quadrant (36.5%)
4. 113 atrial arteries in the left anterior quadrant (30.1%)

The total number of atrial arteries was 125 (33.3%) in the right half and 250 (66.7%) in the left half of the heart, and 192 (51.2%) in the anterior half and 183 (48.8%) in the posterior half of the heart.

Williams et al. (1989) described anterior, lateral, and posterior atrial arteries serving the right atrial walls and anterior, lateral, and posterior atrial arteries for the left atrial walls.

The largest artery in the atrial circulation is the SA node artery; the latter is single and thick. It arose from the proximal portions of the stems of the RCA in 65%, or of the CA in 32%, of the cases examined by von Lüdinghausen (1975), and DiDio (1995).

The AV node artery is also quite strong; in the crux cordis area it derived from the terminal portions of the RCA in 90%, or the CA in 10%, of the cases examined by von Lüdinghausen (1975), and DiDio (1995).

In 22%–48% of cadaver specimens examined by Björk (1966), and Dupont and Riquet (1991) the bronchial arteries were found to release little branchlets, which communicated with atrial branches.

4.5.1 The Arterial Supply of the Sinuatrial Node

The sinuatrial (SA) node enjoys "a special arterial supply" (Keith and Flack 1906; Gross 1921; Kennel and Titus 1972; James 1977) (Table 2). The branch to the SA node is the strongest atrial branch and shows considerable variation in size and length, irrespective of whether it arises from the RCA or from the CA. The mean external diameter of the SA node branch at its origin – measured radiologically – was found by Kennel and Titus (1972), and He et al. (1991) to be 1.7 mm.

Origin of the SA Node Arteries

According to May (1960), Truex et al. (1967), Davies 1971; Kennel and Titus (1972), von Lüdinghausen (1975), James (1977), Anderson et al. (1979), and Kyriakidis et al. (1988), the stout right branch to the SA node originated in about 55%–65% of heart specimens as the first atrial branch from the first or second proximal centimeters of the RCA. The left branch to the SA node derived in 35%–45% from the first or second proximal centimeters of the CA. In one-third of cases it arose from any point along the length, even including the distal portion, of one of the coronary arteries in the posterior coronary sulcus near to the crux cordis. In the area of the latter, the SA node

Table 2. The origin of single and double SA node artery in the literature

Author(s)	Single SA node artery RCA	LCA	Double SA node artery RCA and LCA
Kennel and Titus (1972)	65%	35%	
Waller (1983)	50%	60%	40%–50%
von Lüdinghausen (1975)	65%	32%	2%–3%
James (1977)	50%–52%	41–46%	2%
Hutchinson (1978)	65%	35%	
Hadziselimovic (1982)	60%	40%	
Kyriakidis et al. (1988)	60%	40%	8%
Futami et al. (2002)	77%	23%	

branch may arise as a continuation of a terminal superior branch of the CA (or the distal RCA) and pass over the posterolateral left atrial wall to reach the sinus node region.

In 32% of heart specimens, the artery ramified and then terminated within the node, where its mean external diameter was found to be 0.7 mm by Kennel and Titus (1972). In 52% of cases examined by McAlpine (1975), the branch created by the ramification coursed through the center of the node, continuing distally to the atrial wall.

Course of the SA Node Arteries

In 44% of the cases examined in this study, the right branch to the SA node approached the node from the posterior aspect, circling the posterior atrial wall in an anticlockwise direction (looking down at the heart from above) and forming an incomplete arterial ring. The artery turned onto the lateral atrial wall at a considerably lower level to that of the summit of the right atrial appendage and coursed superiorly as well as anteriorly to reach the "tail" of the node.

In 40%, the artery approached the node from the anterior aspect, circling clockwise and arching over the crest of the right atrial appendage. On reaching the lateral atrial wall the artery coursed postero-inferiorly to reach the "head" of the node.

The SA node branches were found to approach both anteriorly and posteriorly, and anastomose within the substance of the node, in 2%–16% of the cases studied by McAlpine (1975), James (1977), von Lüdinghausen and Valeske (1978), and Anderson et al. (1979). The SA node artery may be part of a complete "sinuauricular" anastomotic ring. Such an arterial ring around the atriocaval junction has been found to be the exception rather than the rule (Hutchinson 1978, Hadziselimovic 1982; Futami et al. 2002).

In a few cases a *dual supply of the SA node* with two SA node arteries has been verified, one arising from the RCA and one from the CA (Anderson et al. 1979).

There is no relationship between coronary dominance and the location of the origin of the SA node artery (von Lüdinghausen 1975; von Lüdinghausen and Valeske 1978).

The mean distances of the origins of the right and left SA node branches from the aortic ostium are 18 mm and 21 mm, respectively (Futami et al. 2002).

Peculiarities

In rare instances the SA node branch originates directly from the right aortic ostium (James 1977; von Lüdinghausen and Ohmachi 2001) (see also Sect. 4.2.6).

Moberg (1968) and Petelenz (1965a) describe a SA node branch deriving from a bronchial artery in a human heart.

Halpern (1955, 1957) demonstrates that the SA node of the rat heart is supplied by a SA node branch which runs from the right internal thoracic artery.

Tanaka et al. (1998) describe a rare variation in which a posterior SA node artery and an accessory AV node artery originated from a common trunk branching from the posterior segment of the CA.

4.5.2 Arterial Supply of the Interatrial and Internodal Pathways

These parts of the conduction system are supplied by left atrial branches deriving from the CA (including the SA node artery) and right atrial branches deriving from the RCA (including the AV node branch). The transatrial anastomotic networks, which function as a bridge between the two coronary arteries situated in the coronary sulci, form the basis for the dual arterial supply of the interatrial and internodal pathways.

4.5.3 Arterial Supply of the Atrioventricular Node

The branch supplying the atrioventricular (AV) node arises from the terminal RCA or initial part of the PIA (Table 3) Though sometimes double, the AV node artery is classically solitary. Arid et al. (2000) named this transient part of the RCA the "principal artery," which – beneath the terminal part of the CS – passes through the crux cordis area. In this area the principal artery describes, in the majority of cases, a U-shaped loop (the "U-turn" of James and Burch 1958; James 1961; Hutchinson 1978) and releases the AV node branch (which is about 25 mm in length) and occasionally one or two companion branches to the AV node area. At the level of Koch's triangle, its course is marked by a close relationship with the zone of attachment of the septal cusp of the right mitral valve on the one hand, and with the orifice of the CS on the other (Sow et al. 1996). In 80%–90% of cases examined by James (1961), Davies (1971),

Table 3. Origin of AV node artery

Author	Solitary A. RCA	Solitary A. CA	Accessory A. RCA or CA	Double A. RCA and CA
James (1961)	85%–91%	4%–13%		2%–4%
Cabrol and Christides (1975)	85%	15%		
von Lüdinghausen (1975)	80%	10%	27% 11%	10%
Hutchinson (1977)	80%	20%	Frequent	
Hadziselimovic (1982)	85%	13%		2%
Sow et al. (1996)	74%	10%		15.8%
Arid et al. (2000)	90%	10%		
Futami et al. (2001)	80%	10%		10%

Cabrol and Christides (1975), von Lüdinghausen (1975), Hutchinson (1978), Arid et al. (2000), and Futami et al. (2001), it derived from the RCA, and in 10%–20% from the dominant CA of the LCA (Fig. 17).

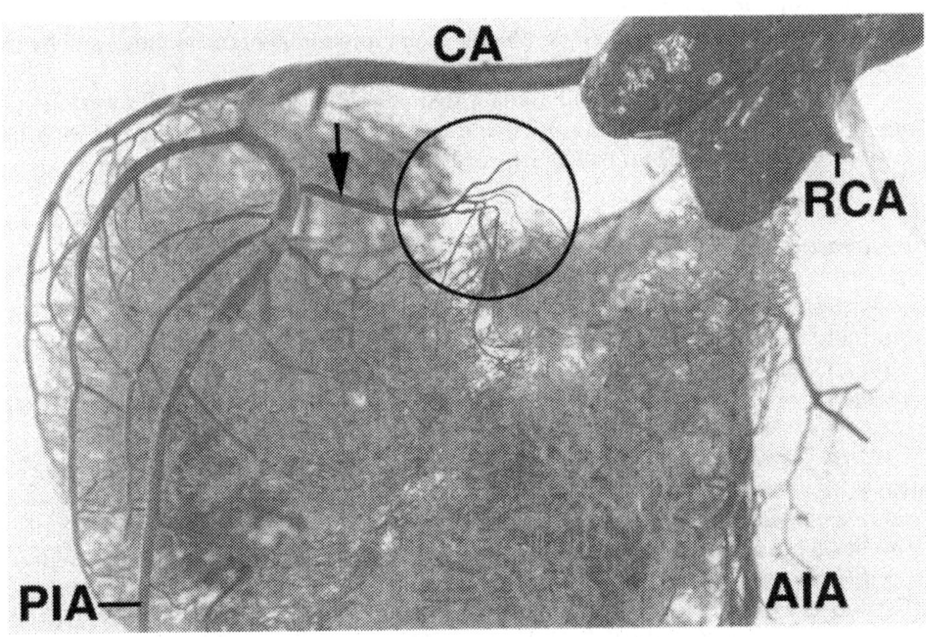

Fig. 17. AV node branch in a corrosion specimen of the LCA (seen from dextro-lateral view). The dominant *CA* releases the AV node branch (arrow) which in turn divides into several branches (indicated by a *circle*). The latter, after describing a right angle, supplies the AV node, AV trunk and proximal left and right bundles

In 38% of cases examined by von Lüdinghausen and Valeske (1978) and in 15.8% examined by Sow et al. (1996), there was an accessory AV node artery arising from the principal artery in the crux cordis; in 10%, a bilateral AV node artery had developed, one branch arising from the RCA and one from the CA.

In some instances the AV node artery is connected with Kugel's artery (Waller and Schlant 1994); in a single case, the AV node artery was found to arise from a posterior SA node artery (Futami et al. 2001).

4.5.4 Arterial Supply of the Common AV Bundle and Proximal Right and Left Bundle

In ten specimens studied by James (1961) and Frink and James (1973), in order to ascertain the supply of the common AV bundle (His) and proximal right and left bundles, *the proximal right bundle branch* was supplied by both the AV node artery and septal branch in five of the hearts, only the septal branch in four, and the AV node artery alone in one.

The distal right bundle branch has its own arterial supply by a strong branch of the AIA: the moderator band artery or descending septal artery (see Sect. 4.6).

The anterior half of the left bundle branch was dually supplied by the AV node artery and septal branch in four of the hearts, entirely by the septal branch in five, and the AV node artery in one.

The posterior half of the left bundle branch was supplied by the AV node artery alone in five of the hearts, dually by the AV artery and septal branch in four, and by the septal branch alone in one.

Thus, the common AV bundle and proximal right and left bundle were dually supplied by the AV node branch and the first anterior IV branch of the AIA in 90%, but entirely by the AV node branch alone in 10% of the cases.

Peculiarities

Ventricular branches: Coronary preponderance effects not only the origin of the AV node and common AV bundle branches but also the arterial branches to the proximal right and left bundles. In rare cases the AV bundle and its right branch is supplied by an anterior septal branch arising from the first antero-lateral (diagonal) artery (Reig et al. 1988, 1989).

Atrial branches: Coronary preponderance also influences the origin of right or left atrioventricular branches. A dominant RCA releases an additional small branch to serve the posterior wall of the left atrium. A normal or dominant CA releases one or two small branches to serve the Koch's triangle and the AV node area or the AV node itself (DiDio and Wakefield 1972; Hutchinson 1978). The CA may give off a posterior "left" artery to the SA node which courses over the posterior wall of the right atrium to reach the sulcus terminalis and the SA node.

4.5.5
The Extracardiac Noncoronary Arteries

Arterial connections between coronary and mediastinal systemic arteries are common and have been visualized many times angiographically, both in patients and in postmortem studies. These extracardiac atrial arteries may arise from bronchial arteries, but also from the right internal thoracic artery, and supply the right atrial region of the mammalian heart (Figs. 18, 19) (Langer 1880; Hudson et al. 1932; Halpern 1957; Baroldi and Scomazzoni 1965; McAlpine 1975; Moberg 1967a,b; Dupont and Riquet 1991). 33% of the patients treated angiographically by Petelenz (1965a,b), and Moberg (1967a,b) and 4% of the anatomical specimens examined by von Lüdinghausen (1978) exhibited one or two branches of mediastinal, mainly bronchial arteries of about 0.5 mm internal diameter coursing over the bare area of the left atrium at the epi-pericardial reflection (Figs. 18, 19). From their ramification pattern it can be said that few of these extracardiac arterial connections represent pure anastomoses; others obviously supply smaller or larger myocardial areas between the pulmonary veins (left sinus venarum).

Tenuous arterial connections between the vasa vasorum of the larger vascular stem from and to the coronary arteries, and Moberg's (1968) case report of the noncoronary supply of the sinus node by an extracardiac artery branching from a mediastinal artery, should be viewed in this context.

Evolutionary and Comparative Anatomy

Extracardiac arteries to the heart in mammals may result from the persistence of a vascular communication within the mediastinal mesoderm in early embryogenesis. In this period of development the heart lies high in the cervical region but later, as a result of differential cephalo-caudal growth, there is an apparent caudal migration to the definitive thoracic position of higher vertebrates. As the heart migrates, the great vessels associated with it are carried caudally in the body during their development (Robb 1965; Baroldi and Scomazzoni 1965; Moore 1988; Angelini 1989).

Therefore, in gill-breathing vertebrates the coronary arteries arise from collecting loops at a position homologous with the upper dorsal aorta. When lungs replace gills, the site of origin of the coronary arteries shifts from the dorsal aorta to the ventral aorta, which is now the nearest supply of oxygenated blood. The site of origin of the coronary arteries then descends from the upper or cephalic portion to the caudal part of the ventral aorta along with the caudal migration of the heart (Halpern and May 1958).

This evolutionary trend has been observed in lower vertebrates: the sites of origin of the coronary arteries migrate from cervical regions down into the mediastinum and almost always closer to the heart. In chondrichthyes and osteichthyes the coronary arteries are branches from the hypobranchial arteries and follow a long, straight course to the heart (Halpern and May 1958). However, in amphibia and reptilia, the sites of origin of the coronary arteries are closer to the absorbed atrial region of the heart, in particular to the sinus venarum (Halpern and May 1958). The ventricle does not receive coronary vessels, but is supplied through its spongy and trabeculated wall (Halpern and May 1958).

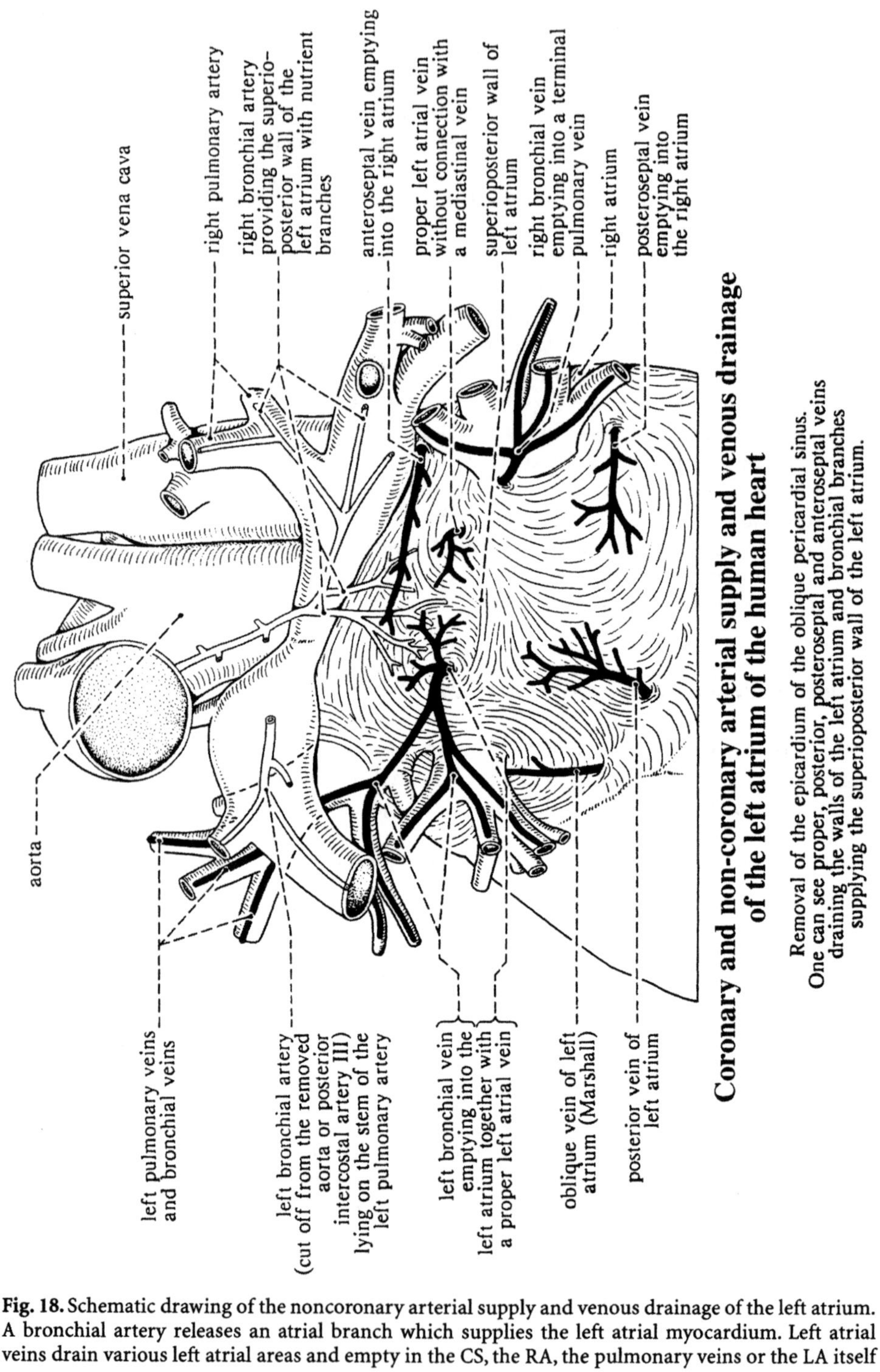

Fig. 18. Schematic drawing of the noncoronary arterial supply and venous drainage of the left atrium. A bronchial artery releases an atrial branch which supplies the left atrial myocardium. Left atrial veins drain various left atrial areas and empty in the CS, the RA, the pulmonary veins or the LA itself

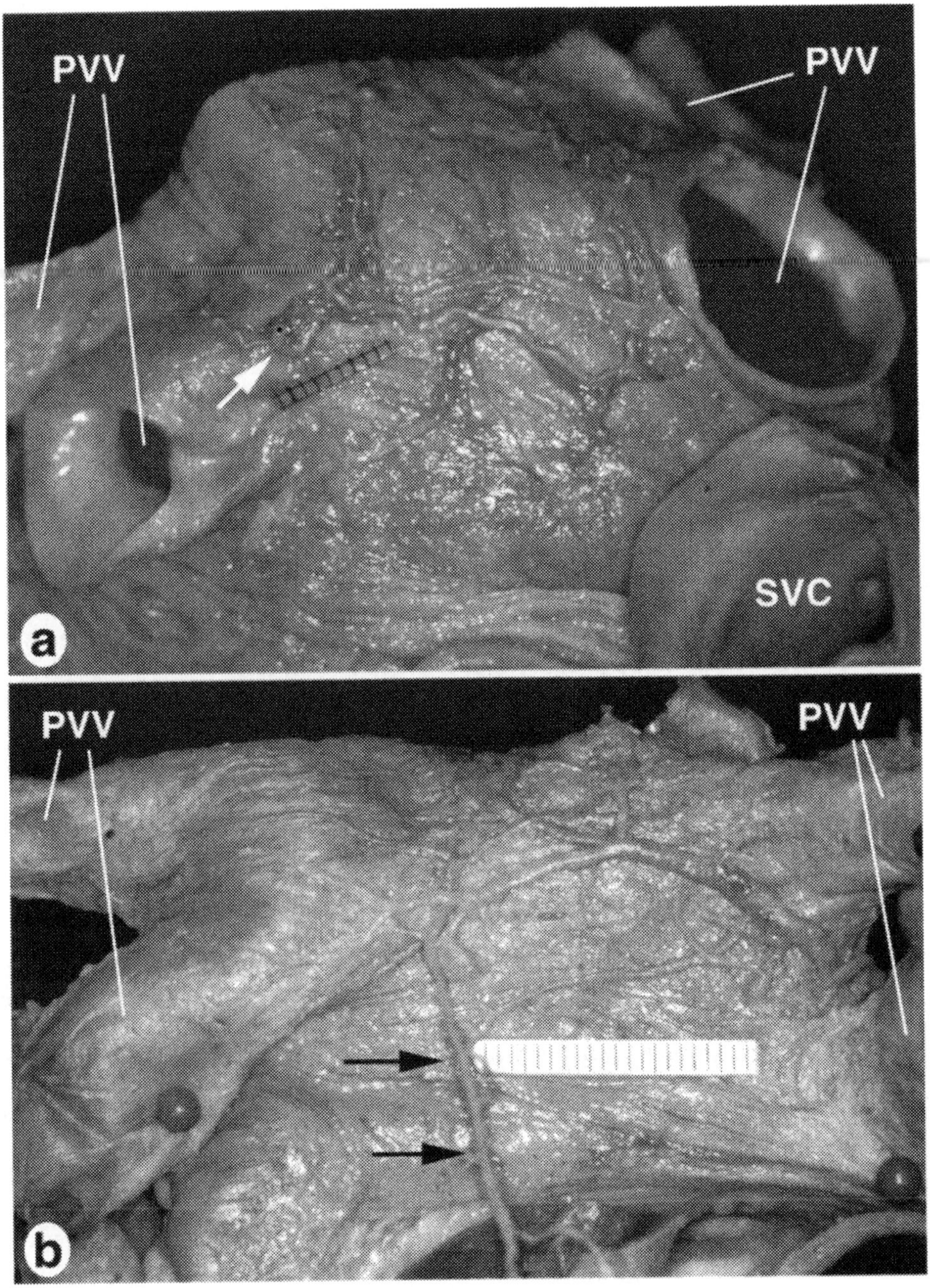

Fig. 19a, b. Superior aspect of the LA (cadaver specimens) showing a noncoronary (mediastinal) artery supplying the myocardium between the openings of the pulmonary veins. **a** The section of the small noncoronary artery is marked by an *arrow*. **b** The extracardiac course of the noncoronary artery is marked by *two arrows*

The hearts of rats and dogs exhibit a dual blood supply: in addition to the coronary arteries there is a considerable number of extracardiac arteries which traverse the bare area and nourish the atrial walls, the sinuatrial nodal tissue and an area of the sinus venarum (Halpern 1954, 1955, 1957; Mizukami et al. 1992).

4.6 Arterial Supply of the Interventricular Septum

The nature of the arterial supply of the interventricular (IV) septum mainly depends upon the length and strength of the AIA and corresponding PIA (Fig. 20).

Usually the septum receives most of its blood supply from the anterior septal branches of the AIA. However, there are two or three other arterial sources for the nourishment of the IV septum (Figs. 21, 22a–c, 23a–d).

In 50% of our cases, the entire septal myocardium could be divided into three territories (von Lüdinghausen and Ohmachi 2001) nourished by: (1) posterior septal branches deriving from the PIA, (2) anterior septal branches deriving from the AIA, and (3) a left superior septal branch deriving from the division of the LCA.

In the other 50% of our cases there were four territories of the IV septum supplied by the above-mentioned branches and additionally by the right superior septal artery (RSSA) deriving from the stem of the RCA. The frequency and distribution patterns of these branches varies conspicuously, as does the size of the territories nourished.

As described above (see Sect. 4.1.2), the AIA reaches the apical region in almost all cases. However, in very rare instances it reaches only the proximal portion of the anterior IV sulcus; here, the distal territory may be supplied by unusually long antero-lateral, diagonal, or right coronary branches (Angelini 1989).

4.6.1 The Anterior Interventricular Septal (Perforator) Branches

In our material, the AIA, and in a few cases its anterior diagonal branches, were found to release 6–8 large anterior septal branches which penetrated the anterior IV septum (Figs. 21, 22a–c, 23a,c). In many cases, a differentiation has to be made between:

1. Septal branches which serve the myocardial layers mainly on the right side of the IV septum (including the moderator band and anterior papillary muscle)
2. Septal branches which serve the myocardium mainly on the left side of the IV septum.

It was observed in group (1) that the first or second anterior septal branch was long and strong and followed a descending, often perpendicular course close to the endocardium on the right side of the IV septum. In the majority of our cases it reached the myocardium of the trabecula septomarginalis (moderator band) and the right anterior papillary muscle and, from there, began its terminal arborization. This artery should be designated as a *left descending septal artery* (*LDSA*) of the AIA [Reig et al. (2000) named it a *moderator band artery*]. It is a strong vessel and was often found to run at a right angles to the third and fourth septal branches and proceeds in direction of the apex cordis.

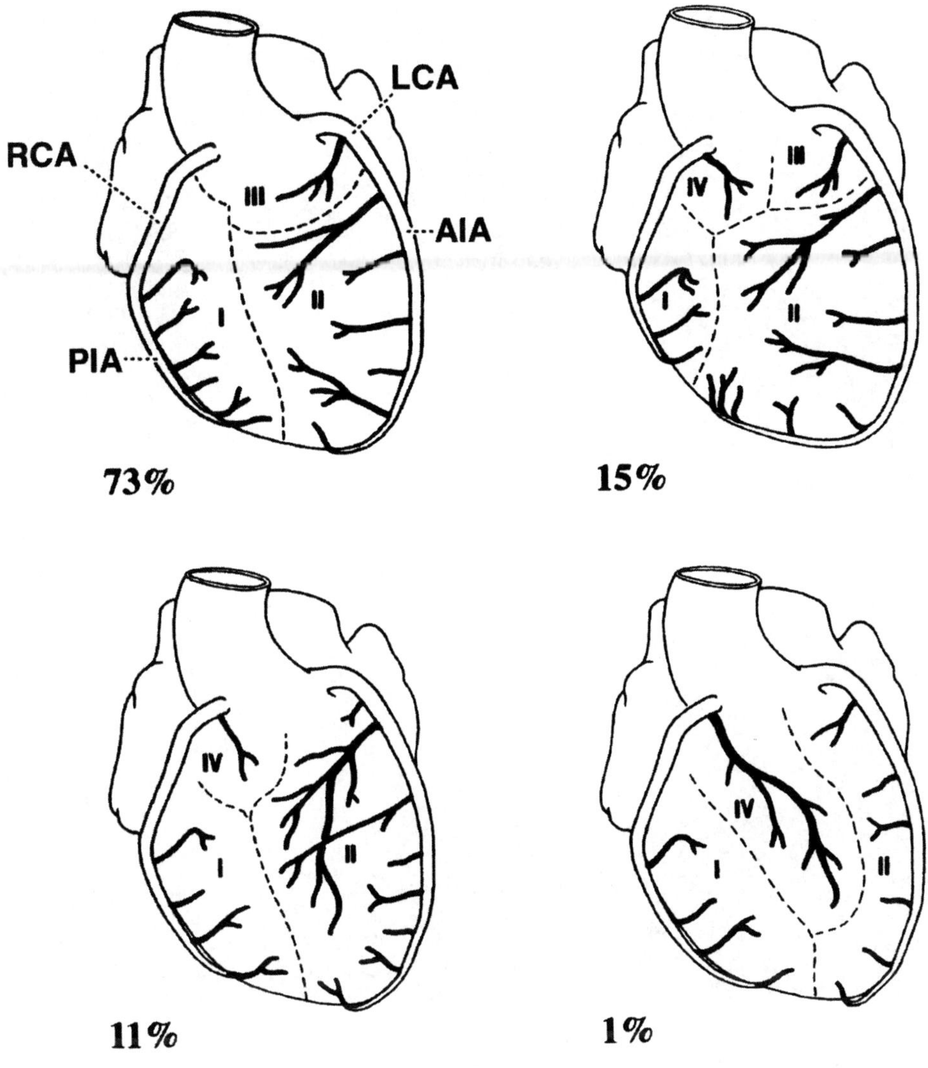

Fig. 20. Schematic drawing showing the frequency and distribution pattern of the arteries supplying the IV septum. Four arterial territories are marked exhibiting: (*I*) posterior septal branches (Rr. septales posteriores of the *PIA*); (II) anterior septal branches (Rr. septales anteriores of the AIA); (*III*) LSSA (R. septalis superior sinister of the *AIA*); (*IV*) RSSA (R. septalis superior dexter of the *RCA*). A dominant LSSA was found in 11% and a dominant RSSA in 1% of our cases

The fourth, fifth, and sixth septal IV branches were smaller and shorter, and arose from the AIA approximately at right angles. These entered, and were subsequently distributed within, the subendocardial layers of the myocardium of the LV.

The length of the anterior branches ranged from 40 to 80 mm (James and Burch 1958; and James 1961). This is in contrast to the posterior septal arteries, which are commonly 15 mm in length.

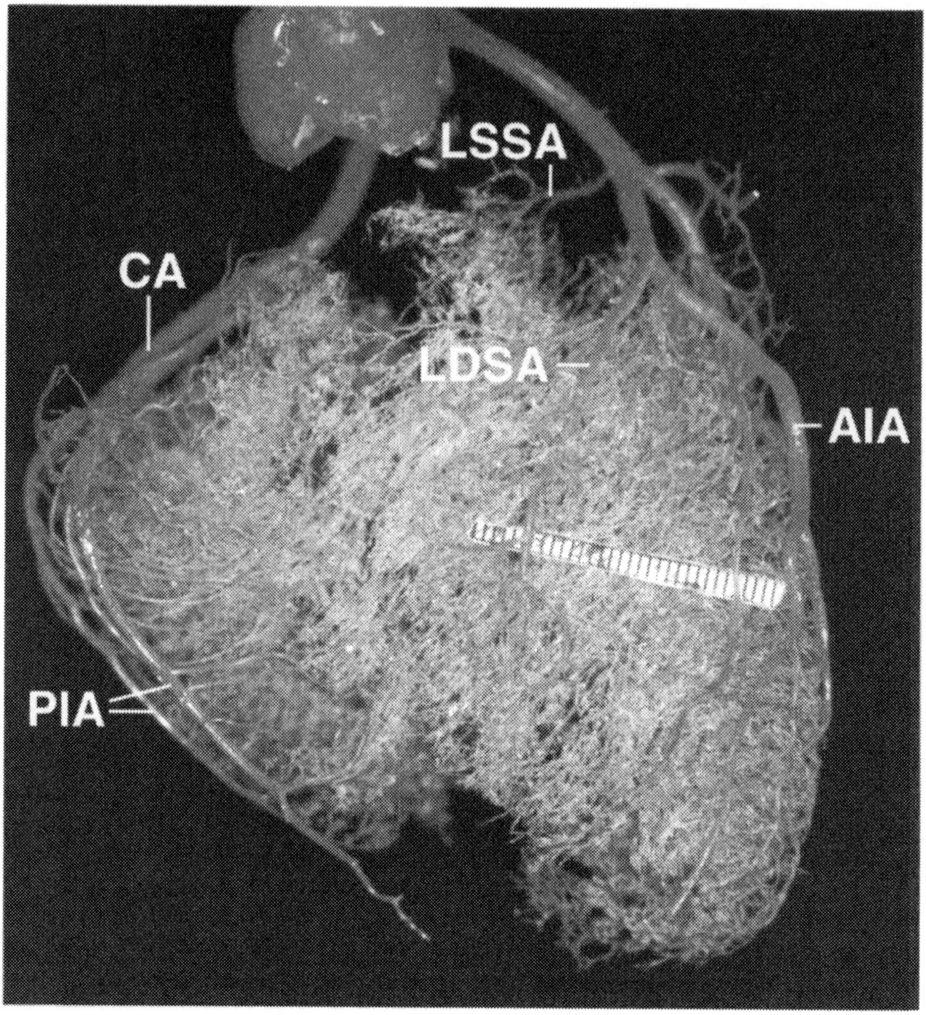

Fig. 21. Arterial supply of the IV septum as seen in a corrosion cast. The proximal *AIA* releases an *LSSA* as a first branch and a strong *LDSA* as a second branch. The *PIA* gives off multiple short posterior septal branches

The exact number and significance of the penetrating branches corresponds to the size and length of the AIA. The septal branches deriving from an AIA which reaches only to the apex cordis nourish only the anterior two-thirds of the IV septum. The septal branches from an AIA which courses around the apex and terminates in the posterior IV sulcus supply the anterior three-fourths of the IV septum (Farrer-Brown and Rowles 1969).

Fig. 22a–c. Corrosion casts showing various arrangements of the *LDSA*. **a** Single *LDSA* supplying a large area of the right side of the IV septum. **b** Duplicate *LDSA*. **c** Triplicate *LDSA*

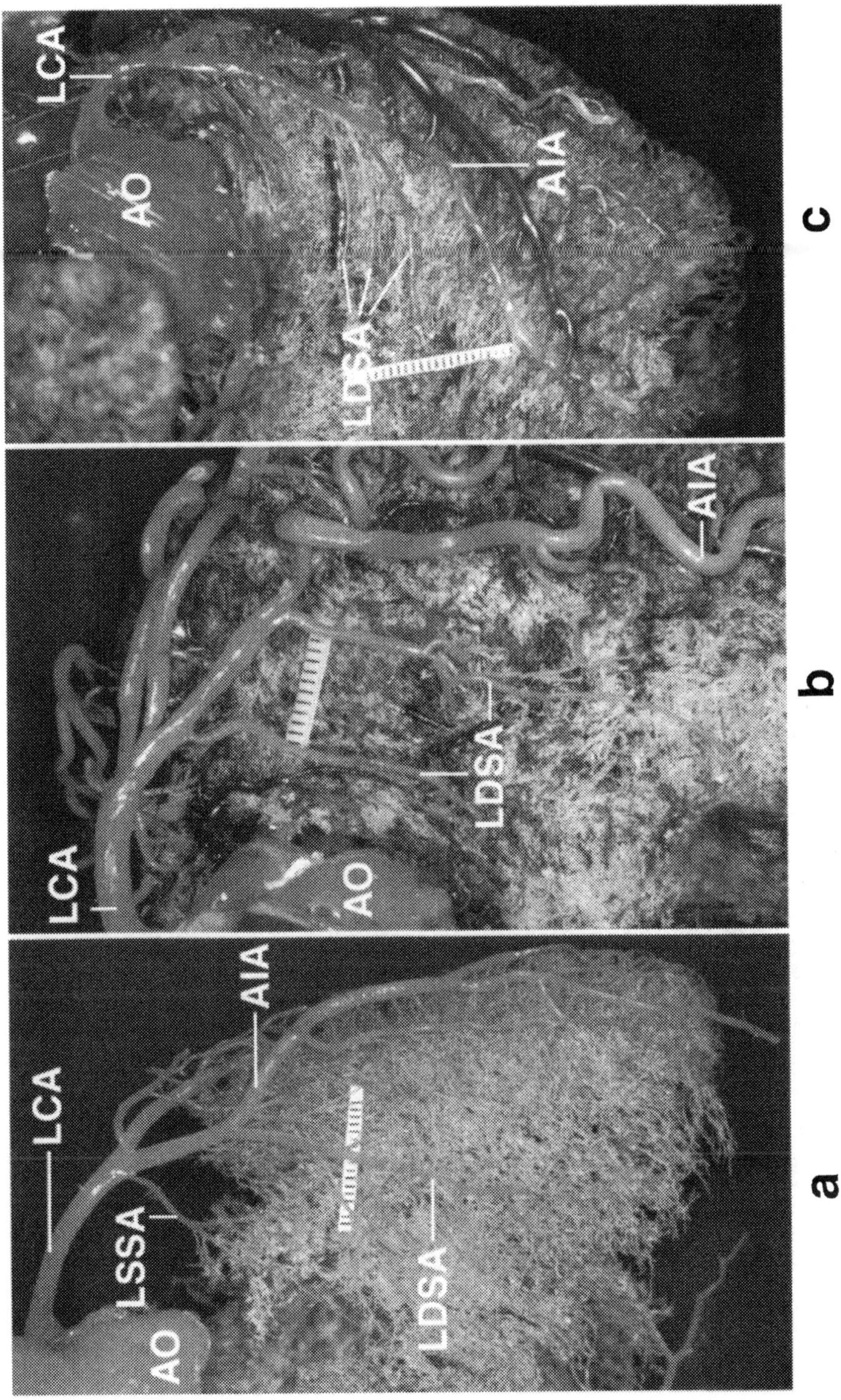

AO
LSSA
LCA
AIA
LDSA
a
LCA
AO
LDSA
LDSA
AIA
b
LCA
AO
AIA
LDSA
c

Superior septal vessels

(right ventricle opened)

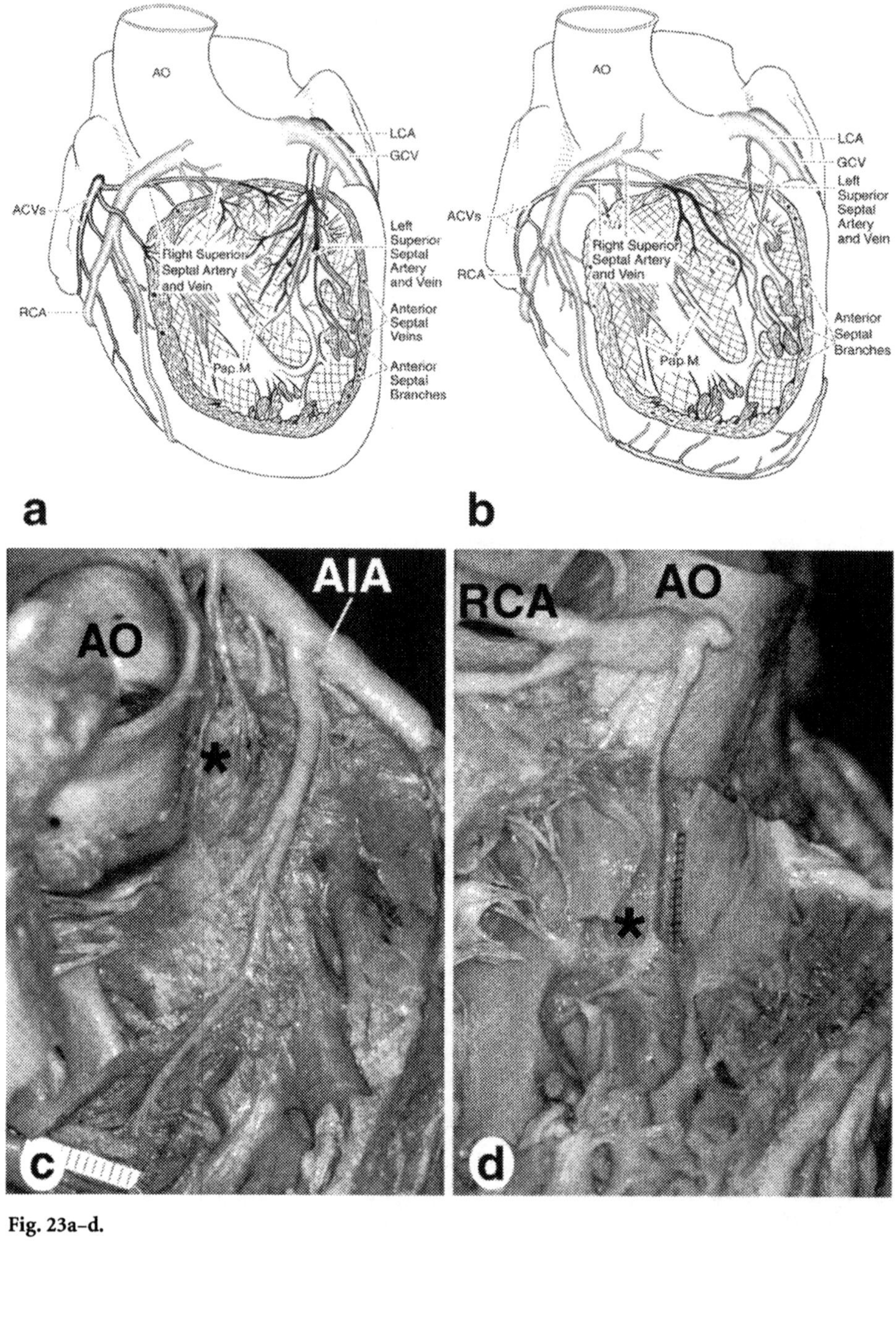

Fig. 23a–d.

Seven (on average) superficial branches are generally found at the level of the anatomical apex. The most frequent occurrence was that of the AIA (91.5%) in the cases examined by Baptista et al. (1988); the frequency of occurrence of the PIA was 14.8% and that of the ramus anterior ventriculi sinistri inferior 11.1%.

4.6.2 The Left Superior Septal Branch or First Anterior Septal Branch

The first septal branch of the AIA was designated the left superior septal branch in 88% of the cases studied by von Lüdinghausen and Ohmachi (2001). However, it was only seen in cast specimens in 12% of the cases examined by Rodriguez et al. (1961) and 9% of the cases examined by Bertho and Gannon (1964). In two of our cases the left superior septal branch (LSSA) arose as the first branch of the CA, and in one case as the first branch of an anterolateral branch of the AIA.

4.6.3 The Posterior Interventricular Septal Branches

The posterior IV septal branches are relatively short, seldom being more than 15 mm in length. There are numerous posterior IV septal branches in proportion to the length of the PIA, but they do not usually reach the apical region (which is supplied by terminal septal branches of the AIA). In 85% of cases examined by James and Burch (1958) and Williams et al. (1989) they were found to nourish about one-third of the IV septum, and in 15% only one-fourth. The largest posterior septal artery (normally the first), commonly originates from the inverted loop of the terminal RCA. It usually represents the AV node branch, and was found to supply the AV node in 80%–90% of cases investigated by Hutchinson (1978) and Williams et al. (1989).

◄

Fig. 23a–d. Arterial supply and venous drainage of the superior part of the IV septum. Schematic drawings of the heart (**a** and **b**) and cadaveric specimens (**c** and **d**) with right ventricles opened to show the main stems of the *RCA* and *LCA* and superior septal vessels. **a** A dominant LSSA (*highlighted*) which, in this case, supplies the superio-anterior part of the IV septum, the trabecula septomarginalis and even the base of the anterior papillary muscle. This artery may be named a *LDSA* or *moderator band artery* (Reig et al. 2000). **b** A dominant RSSA which supplies the myocardium of the superior IV septum and also the trabecula septomarginalis (in contrast to the weak and thin LSSA). **c** Anterior view of the aortic root showing a dominant *AIA* which releases a small LSSA as a first branch (marked by an *asterisk*) and a strong LDSA as a second branch. **d** Anterior view of the aortic root showing the stem of the *RCA* which releases a strong RSSA. After an extramural course, the RSSA ramifies intramurally and supplies the myocardium of the crista supraventricularis and the adjacent IV septum

4.6.4
The Right Superior Septal Branch

A right superior septal branch (RSSA) is defined by its origin from the proximal RCA or right aortic sinus, by an extramural course to the crista supraventricularis, and by intramural branching and its distribution pattern in the myocardium of the crista supraventricularis and superior ventricular septum (Figs. 20, 23b,d).

In 26% of our cases, a small area of the superior IV septum was supplied from one or two tiny RSSA. A RSSA was seen to derive from the first 10 mm of the stem of the RCA in 21% of the cases examined by Campbell (1929), and Bream et al. (1979), from the ostial area of the RCA in 4%, and from the floor of the right aortic sinus in 2% of cases studied by Rath et al. (1986) angiographically and by von Lüdinghausen and Ohmachi (2001) microanatomically.

The RSSA had a length of less than 10 mm and a caliber of less than 1 mm (in 16% of the cases studied). In 9%, the length was 10–17 mm and the caliber about 1 mm.

From its origin, the RSSA courses downwards, giving off occasional branches to the periaortic connective tissue, to the subvalvular myocardium of the crista supraventricularis area, and to the dorsal wall of the conus arteriosus. There is no relation between the frequency of occurrence and coronary dominance.

An RSSA is visualized angiographically in about 3.5% of patients (according to Bream et al. 1979).

Peculiarities

In two cases examined in our study there was a long, strong RSSA of substantial appearance arising from the stem of the RCA. The dissectable lengths of these RSSAs were 35 and 36 mm, and their calibers 1.5 and 1.6 mm. The RSSA entered and nourished the myocardium of the superior third of the IV septum. The corresponding LSSA – stemming off from the first anterior septal branch – was short, thin, and weak (von Lüdinghausen and Ohmachi 2001).

In other cases the RSSA revealed its early ectopic origin, showing one or two accessory vascular ostia, 0.3–0.4 mm in diameter, in the floor of the right aortic sinus. The RSSA themselves were 16 and 17 mm in length and nourished the subvalvular myocardial tissue (von Lüdinghausen and Ohmachi 2001).

The RSSA is constant in cattle and guinea pigs and occasionally seen in dogs (Robb 1965).

4.6.5
Arterial Supply of the Right Bundle and Moderator Band

The artery supplying the right bundle and moderator band derives as a LDSA or moderator band artery (Reig et al. 2000). In 33% of cases examined by James (1961) and Futami et al. (2001), and in 65% of our own material it was found to be the second anterior septal IV artery, and in 34% (own material) the 1st anterior septal artery. The LDSA may occur as a single or double vessel. In rare cases there were three small LDSAs. These branches originated from positions 16–48 mm from the orifice of the

LCA. In rare cases the artery supplying the right bundle and moderator band derives as RSSA from the RCA.

The LDSAs join many strong intra- and intercoronary septal anastomoses (Figs. 20–22a,b, 23a,c).

4.6.6
Arterial Supply of the Papillary Muscles

The papillary muscles, which – according to Ranganathan and Burch (1969) – may exhibit a fingerlike shape, be tethered to the subjacent ventricular wall, or be of an intermediate type, play an important role in the maintenance of the atrioventricular valve efficiency. This has become more evident by the clinical syndrome of papillary muscle dysfunction due to ineffective arterial supply.

Right Ventricle

The *anterior papillary muscle* is supplied predominantly by the LDSA or moderator band artery (Reig et al. 2000), both branches of the AIA, which penetrate it. Occasionally the anterior papillary muscle is supplied by the RSSA (Figs. 21, 22a,b, 23b,d).

The *posterior papillary muscle* receives branches from the PIA, which – according to the prevailing dominance – may be a branch of the RCA or the CA.

The *septal papillary muscles* are nourished by anterior septal IV (perforating) branches of the AIA.

Left Ventricle

The *left anterior (anterolateral) papillary muscle* receives branches from the AIA and either the anterolateral (diagonal) branches or terminal branches of the left marginal branch of the CA (Ranganathan and Burch 1969).

The *left posterior (posteromedial) papillary muscle* receives a variable supply, reflecting the various means by which blood reaches the region of the posterior septal attachment. On the one hand a dominant CA, and on the other a dominant RCA and its PIA, may release branches for the supply of this papillary muscle (Estes et al. 1966b; Silverman and Hurst 1968; Ranganathan and Burch 1969; Reiss and Becker 1981; Flaherty et al. 2001).

When the CA supplies nearly all the diaphragmatic surface of the left ventricle, as in 10% of the cases examined by James (1965), Estes et al. (1966 a), and Flaherty et al. (2001), its branches provide the entire supply for the left posterior papillary muscle. When the RCA extends as far as the margo sinister (in about 20% of human hearts) it supplies all the arteries to the posterior (posteromedial) papillary muscle (Ranganathan and Burch 1969).

Ranganathan and Burch (1969) described two predominant types of arrangement of the arteries in the left ventricular papillary muscles. Papillary muscles that were finger-like received a large central artery which coursed through the entire papillary muscle to its apex. Such papillary muscle arteries had very few or no anastomotic connections with the extrapapillary subendocardial plexus of vessels. The papillary muscles that were completely tethered to the subjacent ventricular wall had a segmental distribution of the long penetrating intramyocardial vessels. The papillary muscles of the intermediate type had a combination of both types of vascular arrangements. Both the intermediate type and the completely tethered papillary muscles with the segmental pattern of arterial distribution had many anastomotic connections with the extrapapillary subendocardial plexus.

4.7 Blood Supply to Coronary Arteries (Vasa Vasorum)

The number of publications dealing with the anatomy of coronary vasa vasorum is rather small. Only few investigators have studied the anatomy of nutrient vascular beds of the human coronary arteries (Johnson and Blake 1968). Vasa vasorum may be demonstrated by injection and stereomicroscopic dissection. Vasa in the adventitia are derived from primary, secondary, and smaller branches of the main coronary arteries and their larger branches. They consist of axially oriented venules and arterioles with frequent ramifications to the adventitial-medial junction. The greater number of vasa (20–40 µm in width) are situated on the external surfaces of the arteries and constitute transverse communications. At the adventitial-medial junction, capillaries (5–7 µm in width) penetrate into the media and form intramural plexuses. These plexuses are also frequently joined by vasa arising directly from the arterial lumen (Johnson and Blake 1968).

Occlusion of the coronary arteries may be effected by spontaneous or progredient hemorrhage of vasa vasorum into their walls (Scotti 1977; Riede et al. 1993).

4.8 Coronary Anastomoses and Collaterals

Numerous vascular connections exist in the epimural circulation; they are even more frequent in the intramural circulation. Connections are known to occur among intramural arteries, between intramural arteries and intramural veins, and among intramural veins themselves. In addition, each of these vascular compartments may communicate with the left ventricular cavity via arterio-sinusoidal and arterio-luminal vessels on the arterial side, and veno-sinusoidal and veno-luminal vessels on the venous side (Paulin 1983) (see also Sects. 1.1 and 4.1).

Given the intramural distribution of the coronary vessels, which run perpendicular to the myocardium in the left ventricle and at an oblique/acute angle to it in the right, it is understandable that anastomoses have been found only occasionally in the subepicardial layers. Conversely, they are frequently found beneath the superficial muscular plane and at greater depth within the heart wall.

For the sake of clarity, distinction must be made between "normal" preexisting *anastomoses* and acquired and significantly widened *collaterals* in hearts which exhibit atherosclerotic damage.

Coronary *anastomoses* are congenitally determined channels which are prominent during fetal life and persist until about the eighth postnatal month, when they diminish in size (Castorina 1957; Helfant and Gorlin 1972). However, the normal hearts of adults still possess conspicuous but tenuous coronary anastomoses (Fig. 24a–c). These are commonly found in the atrial and ventricular septa and the surrounding free ventricular walls, at the crux, and in the free walls of both atria. In adults, the significance of anastomoses is that with them, a potential exists for protecting the myocardium against the deleterious effects of ischemia (Baroldi et al. 1956; Esperenca Pina 1974; Schaper 1974; Vlodaver et al. 1976; Cohen 1985; Allwork 1987). It should thus be stated that the development of wide, efficient coronary collaterals depends upon intramurally preexisting anastomoses of the coronary arteries in the normal heart.

In adjacent organs, anastomoses are also found between the coronary arteries and the systemic arteries, including the vasa vasorum of the aorta, of the superior and inferior venae cavae, and of the pulmonary artery, as well as the pericardial arteries along the reflection of the pericardium (Baroldi and Scomazzoni 1965; Angelini 1989; von Lüdinghausen et al. 1995).

Coronary *collateral* circulation is found both clinically and experimentally in the various severe pathological syndromes. Well-developed collaterals were found angiographically in 75% of the cases with severe obstructive atherosclerotic lesions studied by Barmeyer (1968), Gensini and da Costa (1969), Carroll et al. (1974), Vlodaver et al. (1976), Ensslen et al. (1977) and Cohen (1985); they provide a demonstrable protection against hypoxia and ischemia and can be interpreted as the result of hypertrophic evolution of the preexisting collateral vessels which takes place in every normal heart.

Physiologically, collateral circulation can be defined as the increase of myocardial perfusibility of the blood flow which pursues a "channel" system of vessels alternative to, or developed in substitution for, an occluded major vascular pathway (Bloor and Liebow 1965; Barmeyer 1968; Allwork 1987).

In healthy hearts, coronary arterial anastomoses carry little or no flow. In pathological conditions, anastomotic communications are markedly increased. The most significant collaterals coursing into and through a scarred area are found in cases of coronary obstruction near an infarcted area; these collaterals are the product of a slow evolutionary process involving a single coronary vessel (May 1960; Vlodaver et al. 1976; Ensslen et al. 1977; Waller and Schlant 1994).

Anatomically, four types of collateral routes can be distinguished: endomural, intracoronary (homocoronary), intercoronary (ventricular and atrial), and extracardiac (retrocardiac and transepicardial) (Bloor and Liebow 1965; Bloor 1974; Vlodaver et al. 1976; von Lüdinghausen and Valeske 1978; Cohen 1985). The latter three groups of anastomotic vessels often exhibit a corkscrew pattern. They usually constitute the thinnest anastomoses in normal hearts, but sometimes occur as the larger collateral vessels in atherosclerotic hearts.

The newly formed, so-called scar collaterals constitute a separate and distinctive group. These collaterals emanate from vessels near the infarcted area; they are rather numerous and intertwining in nature and may course into and through the scarred area (May 1960). Anastomoses of the coronary arteries are easy to identify since they are not accompanied by veins (Baroldi et al. 1956).

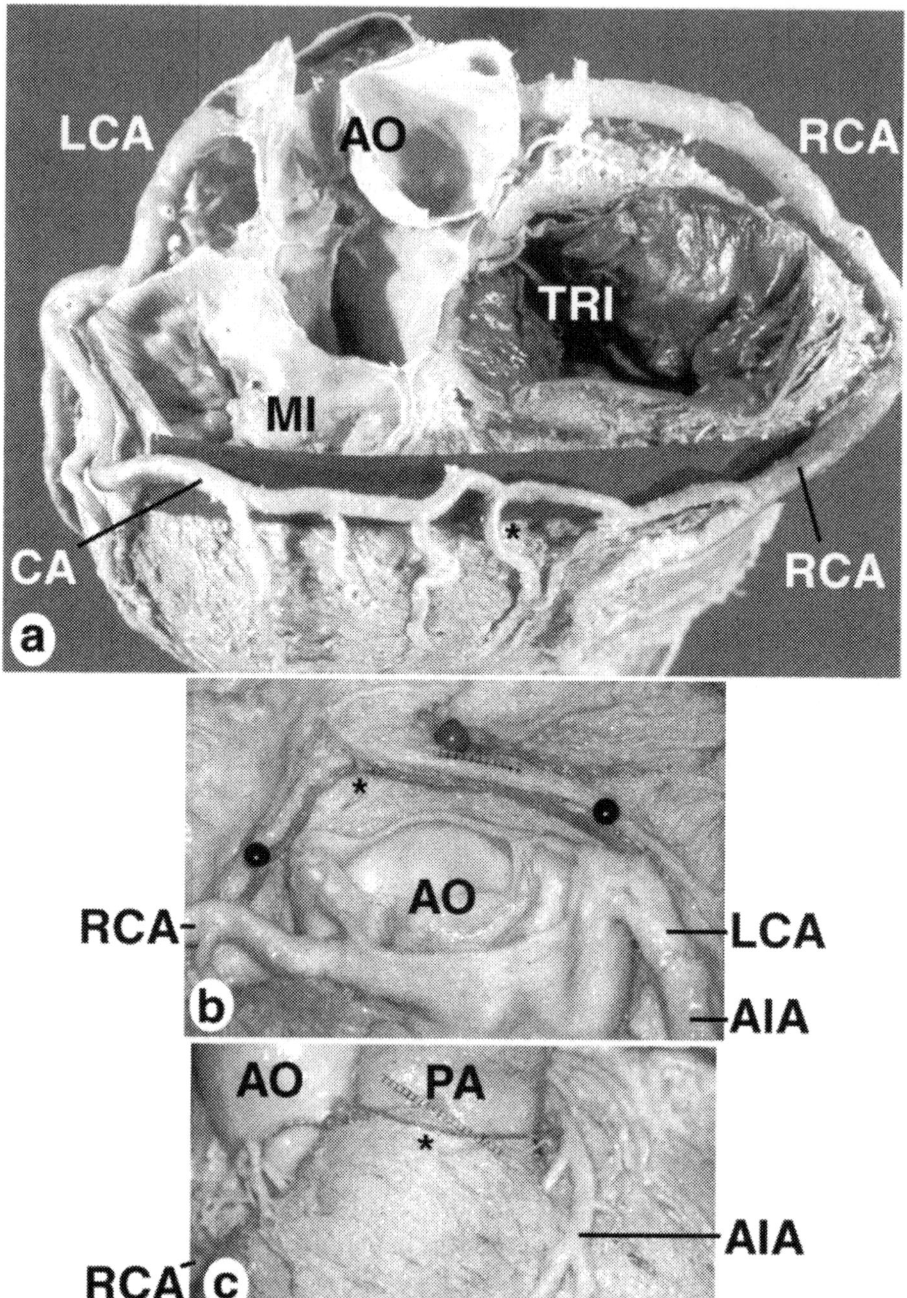

Fig. 24a–c. Intercoronary anastomoses and possible bypass collaterals between the *RCA* and *LCA*. **a** A large anastomosis between the *CA* and *RCA* in the area of the crux cordis (seen from posterior). **b** A retro-aortic anastomosis between the proximal right and left atrial branches (*asterisk*; seen from anterior). **c** A preaortal anastomosis situated on the anterior wall of the root of the pulmonary trunk and connecting the right and left conus branches (seen from anterior)

4.8.1
Anastomoses in the Ventricular Walls

The frequency of intracoronary anastomoses in the myocardium of the LV has been demonstrated by scanning electron microscopy. It has been shown that every arborization of a branch having a lumen of about 500–1000 µm in diameter owns an anastomotic vessel. In the RV, the anastomoses and collateral circulation are much less developed, in proportion to the less abundant vascularization (Baroldi et al. 1956; Lametschwandtner et al. 1990).

4.8.2
Anastomoses in the Atrial Walls

The atrial region with its conduction system contains many intracoronary, intercoronary, and extracoronary anastomoses deriving from three sources: the RCA, the California, and the extracardiac arteries. The most striking anastomoses are those in the region of (1) the SA node and (2) the AV node and atrial septum (see also Sect. 4.8.5; Kennel and Titus 1972; Vlodaver et al. 1976; James 1977; Anderson et al. 1979).

In principle the atrial arteries function (1) as bridges between the two major coronary trunks (the RCA and the CA) which lie in the coronary sulcus, and (2) as bridges and collaterals between coronary and extracardiac, noncoronary (mediastinal) arteries (Halpern 1954; May 1960; James 1977).

The region of the SA node is supplied by the strongest atrial branches deriving either from the first few centimeters of the RCA or the CA. The SA node branch anastomoses with the right and left intermediate branches of the RCA and posterior left atrial branches of the CA. Communication and collaterals with extracardiac arteries – chiefly bronchial arteries – have been reported by Petelenz (1965a,b), Bloor and Liebow (1965), Moberg (1967a,b), McAlpine (1975) (see Sect. 4.5.1).

The region of the AV node and the lower portion of the atrial septum are supplied from branches which arise from the terminal RCA, from the initial PIA, and from the CA. The AV-node artery, with its accompanying and accessory branches, is usually the longest and strongest artery of the crux cordis region.

4.8.3
Endomural (Subendocardial) Anastomoses

The designation "endomural anastomoses" is used for the precapillary anastomoses in the subendocardium of the left ventricular walls and IV septum. They belong therefore to the noncoronary vascular system (Bloor and Liebow 1965). Under normal conditions, this system has a function other than that of providing collaterals. It consists of channels (50–250 µm in diameter) connecting intramural arterioles and venules, i.e., the vessels of the smallest cardiac vascular system, among them the arterio-luminal and arterio-sinusoidal channels and veno-luminal and veno-sinusoidal vessels, with the chambers of the heart (Wearn et al. 1933; Moir et al. 1964; Fulton 1965).

The tributaries of this system do not primarily act as bypass vessels but may in special cases enable reverse blood-flow from a heart chamber into the myocardium and therefore replace coronary blood flow in an ischemic area threatened by infarction. Some authors are strongly opposed to this assumption, arguing that the subendocardial layers, instead of constituting the myocardial region most vulnerable to ischemic damage, might be the best protected (Griggs et al. 1972; Allwork 1987).

4.8.4
Intracoronary (Homocoronary) Anastomoses

Connections between various branches of the same artery are named intracoronary or homocoronary anastomoses. In the human, these occur most commonly in the IV septum, crista supraventricularis, and moderator band (septal anastomoses can only be designated intracoronary when the AIA and PIA are both branches of the LCA, as seen in 10% of our cases (see Sect. 4.3.3) and in the interatrial septum, but they have also been reported in the ventricular walls near the interventricular sulci, the apex cordis, and the crux and the walls of both atria (James and Burch 1958; Baroldi and Scomazzoni 1965; Farrer-Brown and Rowles 1969; Kalbfleisch 1975; Reig-Vilallonga et al. 1987; Reig et al. 2000).

In adults such anastomoses were frequently found intramurally, below the subepicardial layers, with diameters ranging from 70–300 μm by Belman and Frank (1958) and Baroldi and Scomazzoni (1965); anastomoses up to 80 μm in diameter were found by Bloor and Liebow (1965) in infant hearts which had not been subjected to stimuli of anoxia, anemia, or cardiac hypertrophy.

4.8.5
Intercoronary Anastomoses and Collaterals

Connecting branches of two different coronary arteries are designated intercoronary anastomoses. They are found in the border regions supplied by branches from both coronary arteries (and in half of the cases by branches of the conus branch, the so-called third coronary artery). Primarily their location is intramural, in the walls of the ventricles and in the walls of the atria (James and Burch 1958; Pitt 1959; Bloor and Liebow 1965; Schaper et al. 1969; Vlodaver et al. 1976; Cohen 1985) (Fig. 24a–c).

Pepler and Meyer (1960) investigated heart specimens and coronary arteries which were macroscopically completely free from atherosclerosis: the authors found a consistently higher frequency of interarterial coronary anastomoses in Bantu hearts than in European hearts.

Widened, tortuously winding collaterals, the largest with a caliber of about 350 μm, were most frequently found angiographically by Farrer-Brown and Rowles (1969) in patients exhibiting severe coronary atherosclerosis, and were preferentially distributed in the areas where a dense network of anastomoses is already known to exist:

1. In the atrial walls, where there is an anastomotic and collateral system which connects the atrial branches of the two coronary arteries.
2. In the IV septum of adult hearts, particularly in the middle portion and in the crista supraventricularis where, according to Prinzmetal et al. (1947), Baroldi et al. (1956),

James (1961), and Kalbfleisch (1975), there are on average 20–30 transseptal anastomoses with calibers of up to 300 μm.

According to Reig-Vilallonga et al. (1987) and Reig et al. (2000), the LDSA may play a key role in collateral circulation following obstruction of the proximal RCA, the PIA or AIA:

1. In the upper portion of the IV septum, where large LSSA and RSSA may exist and function as collateral routes to branches of the LCA and even the AV node artery (Campbell 1929; von Lüdinghausen and Ohmachi 2001)
2. In the area of the crux cordis (in the "waist" of the heart), where there are anastomoses and collaterals with calibers of up to 1.5 mm between the RCA and the CA in the posterior coronary sulcus
3. In the apical region, where there are subepicardial anastomoses with calibers of up to 1 mm between the AIA and the PIA
4. Behind the aorta and pulmonary trunk, where there are anastomoses which may connect the stem of the LCA and the initial part of the RCA
5. On the surface of the conus arteriosus, where there are interconnections between the right and left conus branches (ring or circle of Vieussens) (Rathor et al. 1973)

4.8.6 Noncoronary Extracardiac Anastomoses to Atrial Arteries

For better understanding and further clarification differentiation has to be made between (1) atrial branches which serve as bridges with mediastinal (bronchial, internal thoracic intercostal and pericardial) arteries and (2) atrial branches from extracardiac (mediastinal) sources which supply the atrial myocardium (Bloor and Liebow 1965; Björk 1966; Mizukami et al. 1996) (Figs. 18, 19).

These vessels are either remnants of the great retro- and supracardiac mediastinal vascular plexus or derived from the vasa vasorum of the great arteries and veins (Hudson et al. 1932; Bloor and Liebow 1965). According to Hudson et al. (1932), Petelenz (1965a,b) and von Lüdinghausen et al. (1995) the noncoronary extracardiac anastomoses arise as tortuously winding vessels from the mediastinal (pericardial and bronchial) vessels and enter the heart through the pericardial reflections surrounding the pulmonary and systemic veins.

In rare cases (both in the adult and new-born human) an accessory supply to the right atrium is derived from the pericardiaco-phrenic branches of the internal thoracic artery or from the subclavian artery (Halpern 1954; Halpern and May 1958; Björk 1969).

Moberg (1968) found a single case of a strong, dissectable SA node artery arising from a right bronchial artery. Moreover, Moberg (1968) and von Lüdinghausen et al. (1995) described the arterial supply of the left sinus venarum by branches deriving from mediastinal arteries which crossed the bare area within the epi-pericardial reflection. It is assumed that these vessels also served as an extracardiac anastomotic route.

4.8.7
The Significance and Evaluation of Anatomically or Angiographically Demonstrated Anastomoses and Collaterals

Coronary catheterization and angiography enable the clinician to demonstrate the development, localization and quality of coronary anastomoses and collateral vessels; their presence may be indicative of severe and extensive obstructive coronary artery disease. The maximum increase in the number of collateral vessels was observed by Baroldi et al. (1956) and Vlodaver et al. (1976) in cases of coronary narrowing; a less-pronounced increase was found in conditions of chronic hypoxemia.

The normal human heart possesses arterial anastomoses with a luminal diameter of 20–350 μm and lengths ranging from 1 or 2 mm to 4 or 5 mm, according to Baroldi et al. (1956) and Piek and Becker (1989) who studied postmortem angiograms.

Obstruction of the Proximal RCA

Occlusion of the proximal right coronary artery (RCA) may cause the development of collaterals from the conus branch, from the branches of the AIA and the CA (atrial collaterals) (Rathor et al. 1973).

Obstruction of the Distal RCA

Occlusion of the distal RCA may cause the development of septal collaterals from the AIA.

Occlusion of the Distal RCA and AIA

In such a case the collaterals to the occluded vessels may derive from intramural branches in the free ventricular walls.

Occlusion of the RCA and AIA

The major collateral flow to the occluded vessels derives from the CA and atrial branches.

Obstruction of the AIA

This kind of occlusion may cause the development of dual collateral supply from the conus branch, from muscular branches of the RCA, from intramural branches of the CA, and (in 10% of cases generally which exhibit left dominance) through septal branches of the PIA.

Obstruction of the CA

In instances of the obstruction of the CA, its distal segment may be supplied by a vessel which has followed a course through the atrial septum. In accordance with the description offered by Kugel (1927), this artery has been named the Ramus atrialis anastomoticus or – for short – Kugel's artery. It arises only rarely from the LCA, occasionally from the RCA, and more commonly from the CA. These findings have been confirmed anatomically by Gross and Kugel (1934), McAlpine (1975) and Hadziselimovic (1982) and angiographically by Vlodaver et al. (1976), who emphasized that unobstructed angiographic demonstration of the normally minute Kugel's artery is extremely rare.

A strong Kugel's artery effecting the perfusion of the distal segment of the CA can be variously situated in the interatrial septum:

1. Posteriorly, to effect an anastomosis at the crux
2. Inferiorly, to anastomose with a low right anterior atrial artery
3. Posteriorly, to anastomose with a branch from the latter

Hadziselimovic (1982) named as a Kugel's artery not only a connection in the lower portion of the atrial septum, but any strong and dissectable anastomosis between the right and left SA node arteries that had undergone simultaneous development; he found such interconnections in up to 65% of his anatomical specimens.

The functional role of the human coronary collateral circulation remains controversial. Obviously, the mere presence of such anastomotic channels is insufficient to prevent occurrence of infarction. On the other hand, experimental data have shown that the size and spread of infarcts may depend on the presence or absence of a well-developed collateral network (Piek and Becker 1989).

Therefore, angiographical demonstration of collaterals to the coronary arteries is of little clinical importance since it provides no indication of the physical state of the patient or of the functional state of the myocardium (Björk 1969). This is in agreement with the findings of Baroldi (1956), whose postmortem studies of injected human hearts with coronary artery disease established no definite correlation between the number of collateral vessels and the extent and degree of coronary stenosis.

The number and size of extracardiac coronary collaterals increases with age, but it is not clear whether this is independent of atherosclerosis (Moberg 1967a,b; Kitzman and Edwards 1990). Dissectable connections have been found between epimurally coursing arteries in many anatomical specimens from octogenarians and nonagenarians. These connections of up to 1.5 μm thickness existed between the AIA and PIA, between the right and left conus branch, between the right and left SA node arteries, and between the CA and the RCA in the crux cordis area, in specimens both with or without atherosclerosis. They have a true anastomotic function, but their importance for the myocardial blood supply cannot be evaluated. Anastomoses and collaterals are considered to be functionally effective when they are found through angiographic demonstration in the peripheral portion of an artery distal to a severe stenosis or occlusion of it (Björk 1969).

4.8.8
Collaterals in Patients with Ostial and Coronary Spasm

Under unusual circumstances collaterals between the main coronary arteries may have developed in the absence of evidenced vascular disease. This may be the case in patients who exhibit transient spasm associated with only mild atherosclerosis of the coronary arteries, as occurs in young subjects; this is the so-called Prinzmetal's variant of angina pectoris (Prinzmetal et al. 1959, 1960; Gianelly et al. 1968; Maseri et al. 1975; James 1980).

A "high take-off" position of the ostium of a coronary artery may cause long-term spasm of the aortic orifice (Vlodaver et al. 1976). Occasionally coronary catheterization maneuvers cause (transient) coronary ostial spasm, ischemia, and infarction; the spasm may cause the development of collaterals (Vlodaver et al. 1976).

4.9
The Coincidence of Coronary Artery Anomalies and Atherosclerosis

The combination of normal and anomalous coronary arteries is indicative, in adults, of pending atherosclerosis (Liberthson et al. 1979; Keren et al. 1982). However, there is still controversy over whether or not coronary anomalies favor the development of atherosclerosis. No significant difference was found between groups of specimens with and without anomalous arteries with respect to the degree of atherosclerosis or the incidence of myocardial infarction. Where mechanical factors are concerned, whether or not the anomalous artery originates from the aorta or the pulmonary trunk is of crucial importance.

4.9.1
Anomalous Coronary Arteries with Aortic Origin

When the LCA originates from the right aortic sinus its branches, i.e., the AIA and CA, as well as the "normal" RCA, exhibit significant atherosclerotic narrowing (Keren et al. 1982).

When the CA originates as a branch of the RCA or arises from a separate ostium in the right aortic sinus (Click et al. 1989) or when there is an ectopic origin or "high take-off" of one coronary artery (see Sect. 4.2.7) atherosclerosis will develop to the same degree as in normally distributed coronary arteries. By contrast, anomalously originating and coursing coronary vessels develop a significantly greater degree of stenosis than that in nonanomalous arteries in age- and gender-matched control patients (Bader 1963; Spring and Thomsen 1973; Keren et al. 1982; Click et al. (1989).

There are a number of reported instances of "a single coronary artery" in individuals who remain symptomless; in others, severe atherosclerosis may develop, and angina, myocardial infarction, or sudden death may be caused. This constellation, after timely diagnosis, requires surgical intervention for the improvement of the nutrition of the left ventricular myocardium (Alexander and Griffith 1956; Stroobandt et al. 1974; Sharbaugh and White 1974; Kafrouni et al. 1981).

There is a rather frequent occurrence of angina pectoris and atherosclerotic lesions in patients with congenital coronary *arterio-venous fistula*. In these patients relatively small shunt flow and "coronary steal" may exacerbate angina pectoris (Liberthson et al. 1979).

4.9.2
Anomalous Coronary Arteries with Pulmonary Trunk Origin

Cases of the anomalous origin of one coronary artery from the pulmonary trunk underline the importance of physical factors in the development of atherosclerotic lesions: the single "normal" originating artery had developed plaques, stenoses, and even occlusions, while the anomalous vessel appeared free of atherosclerotic damage in specimens studied by Alexander and Griffith (1956), Ogden (1970), and Baroldi (1974). In such a case, wherever – after coronary surgery – there is a change in flow distribution in the anomalous artery (that is to say, "arteriolization" of the anomalous vessel) some modifications of the vessel wall can be expected which may ultimately result in a progressive, often obstructive atherosclerotic lesion. Similar observations were made by Baroldi (1974) over a relatively short period of time in saphenous vein grafts which bypassed severe coronary stenoses.

4.10
Age-Related and Atheromatous Coronary Calcifications

4.10.1
Coronary Calcifications

May (1960) and Möttönen (1969) considered calcification a terminal phase of an atheromatous process. Blumenthal et al. (1944) suggested, however, that it may also play an important role in the initiation of atheromata.

Coronary calcifications or calcium scores in atheromata or just located intramurally were found by Frinck et al.(1970) in 69% of the hearts from randomly selected subjects from the autopsy room, aged 30 years and more at death, and by Pyke and Symons (1951) and von Lüdinghausen and Valeske (1978) in 60%–84% of the hearts from randomly selected subjects, aged 60 years and more at death.

Unselected heart specimens obtained during pathological examinations from individuals aged between 30 and 69 years at death showed that 93% of all coronary arteries with stenosis of 50%–100% and 20% of all coronary arteries with stenosis of 0%–50% contained calcification deposits (Mautner et al. 1994).

Calcifications are most frequently found in the bifurcation of the main LCA and in the initial section of the AIA in the human, regardless of age, sex, and presence of cardiac disease (Montenegro and Eggen 1968; Frink et al. 1970; Grottum et al. 1983).

The pattern of distribution of calcification along the length of the coronary arteries is characteristic and similar, in both sexes, for each coronary artery (Montenegro and Eggen 1968; von Lüdinghausen and Valeske 1978).

4.10.2
Age-Related (Nonatheromatous) Calcifications in the Coronary Arteries

Nonatheromatous calcifications are considered to be age-related. These dense calcifications develop either singly or numerously as small, discrete, stippled foci with a maximum diameter of 1–2 mm. The alterations are established within the subintimal layers, and therefore in preexisting tissue components (not in recently acquired atheromatous plaques); because the calcified foci are not related to atheromata and narrowings the subjects are almost asymptomatic. A calcification deposit may cause the thin media to bulge outwards into the adventitia (von Lüdinghausen and Valeske 1978; Mautner et al. 1994).

4.10.3
Atheromatous Calcifications in the Coronary Arteries

From the clinical viewpoint the predominant cause of coronary artery calcification is atherosclerosis in relatively young individuals (Montenegro and Eggen 1968; Grottum et al. 1983; Fallavollita et al. 1994; Mautner et al. 1994). According to Bakst et al. (1974), Bierner et al. (1978), Crochet et al. (1981), Baron (1994), Davies (1994) and Fallavollita et al. (1994), the calcified masses of more than 2 mm in size which were observed in almost every patient suffering from moderate or severe coronary heart disease had developed in atheromata (plaques) and narrowings which closed/occluded more than half of the diameter of the artery. Their findings indicate the presence of partially obstructive atherosclerosis (in 60%–84% of the patients suffering from coronary heart disease who were investigated angiographically).

Because atherosclerosis is a ubiquitous disease unrelated to old age, it is rare to find a person in the fourth decade of life or older without anatomical evidence of atheromatous disease in the coronary arteries. One third of these relatively young patients, when undergoing coronary catheterization, exhibit atheromatous calcification (calcific atheroma, plaque, or fatty streak) composed of a lipid core or extracellular lipid pool, with numerous lipid-containing macrophages and smooth muscle cells surrounding it. The atheromatous calcification is usually covered by a fibromuscular capsule of varying thickness that separates the blood flow from its contents (Vlodaver 1976; Deedwania and Carbajhal 1991; Baron 1994). Depending on the size of the atheroma and its calcification deposit, and the degree to which the latter is surrounded by connective tissue, the tunica media is significantly atrophic.

Many authors have used combined fluoroscopical, tomographical, angiographical and histological methods (DiDio and Wakefield 1972; Baroldi 1974; Bierner et al. 1978; Koischwitz et al. 1982) to establish the significance of severe atherosclerotic lesions in the coronary arteries.

The degree of severity of the atherosclerosis (with its stenoses and occlusions) is in direct proportion to the site and extent of severe atheromatous calcification in the walls of the two coronary arteries (Bierner et al. 1978; Grottum et al.1983; Baron 1994; Haberl 1998; Knez 2000). Calcific atheromata are more frequently found in patients exhibiting a short main LCA or anomalies of the coronary arteries (Gazetopoulos et al. 1976) and less frequently when there are long intramural courses (Fulton 1965; von Lüdinghausen and Valeske 1978).

The SA node artery is, for unknown reasons, always found to be free of atherosclerotic alterations (Kitzman and Edwards 1990).

In conclusion, an atheromatous calcification will develop into, or at, a mature subintimal plaque, in most cases related to subintimal hemorrhage, stenosis or even occlusion (Fallavollita et al. 1994). According to the findings of Mautner et al. (1994) calcification deposits seem to be an indicator of the location of atheromatous plaques.

4.10.4
Coronary Calcifications in Heart Specimens of Subjects from the Dissection Room

Our specimens obtained from subjects studied in the dissection room constitute a representative sample of members of the population aged between 64 and 92 at the time of their death. In such an unselected and nonhomogeneous group, the heart specimens and their coronary arteries exhibit both atheromatous and nonatheromatous alterations. The vessels often follow tortuous courses, and are thickened and hardened. In any study of the coronary arteries in anatomical heart specimens it is very difficult to differentiate unequivocally between "purely" age-related alterations and "purely" atheromatous lesions (Schlesinger 1941; von Lüdinghausen and Valeske 1978; Kitzman and Edwards 1990); age-related coronary alterations and obstructive atherosclerotic lesions coincide and fuse together (Noltenius et al. 1976; von Lüdinghausen and Valeske 1978, 1979; Crochet et al. 1981; Davies 1994) (Figs. 25a,b, 26a–c, 27a–c).

4.10.5
Evaluation and Measurement of Calcifications

In a postmortem X-ray and histomorphometric study (n=100; subjects aged 60 years or more at death; (von Lüdinghausen and Valeske 1978, 1979), the calcifications of the coronary arteries were localized, evaluated, and classified semiquantitatively. They were designated as mild, moderate or severe according to the few, multiple or numerous foci observed.

Three types of focus were identified, based on size (type A is small, up to 2 mm in size; type B is a medium-sized focus up to 4 mm in size; type C is a large focus more than 4 mm in size) (Fig. 26a–c).

In the 100 cases studied the overall incidence of coronary calcifications was 84%. The RCA was involved in 72% and the LCA in 84% of the cases, and the main stem of the LCA only in 16% of the cases. The quality and amount of the calcification deposits was the same in the three coronary arteries.

The mild type of focus, type A, was found in 19%, the moderate type B in 32%, and the severe type C in 33% of the cases. Type C included a few very long stripes (up to 30 mm in length) and circular or even tubular-developed calcifications.

These stripes and tubular-developed calcifications were found in one coronary artery in 20%, in two arteries in 12%, and in three coronary arteries in 66% of the cases studied.

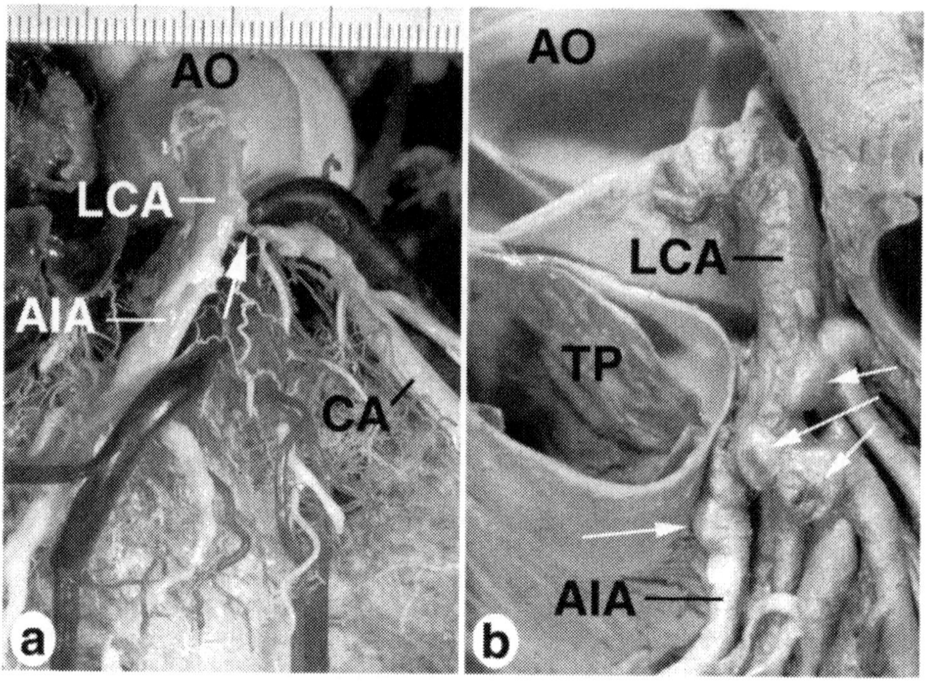

Fig. 25a, b. Arteriosclerosis of the *LCA* as seen in a lateral view (**a** corrosion cast, **b** cadaveric specimen). The initial part of the *CA* exhibits a severe degree of stenosis (*white arrow*). After the division of the main stem of the *LCA* the three branches exhibit four aneurysmata (marked with *white arrows*). The left auricle has been cut away

In specimens with left dominance, the RCA was seen to exhibit calcifications in 4 instances, and the LCA in 11. In specimens with balanced dominance, the RCA exhibited calcifications in 53 instances and the LCA in 68; in specimens with right dominance, calcifications were present in the RCA in 38, and in the LCA in 31 instances.

Most of the calcific deposits were located in the proximal portions (within the first 5 cm) of the coronary arteries.

In 40% of the cases there were 1–4 small foci at a distance of 2.5–3 cm from the aortic ostium. Where the LCA is concerned, this means that the first (two) centimeters of the main artery were free of calcification. In the case of the RCA, it means that the origin of the SA node branch – in most cases – was not hampered by any narrowing through calcification.

In 30% of the cases more, larger calcific accumulations of the same character had developed at a distance of 1.5–5 cm from the aortic ostium. The calcific deposits seemed to have "augmented in number" and "enlarged in size" over time.

Only a few clinicians emphasize that single and small calcium scores are generally not associated with evidence of impaired coronary function and are not indicative of a stenosis at the site of calcification (Pyke and Symons 1951). Other cardiologists assume multiple and large calcifications to be an indicator of the presence of atherosclerotic plaques and to have a strong association and an ancillary role in the

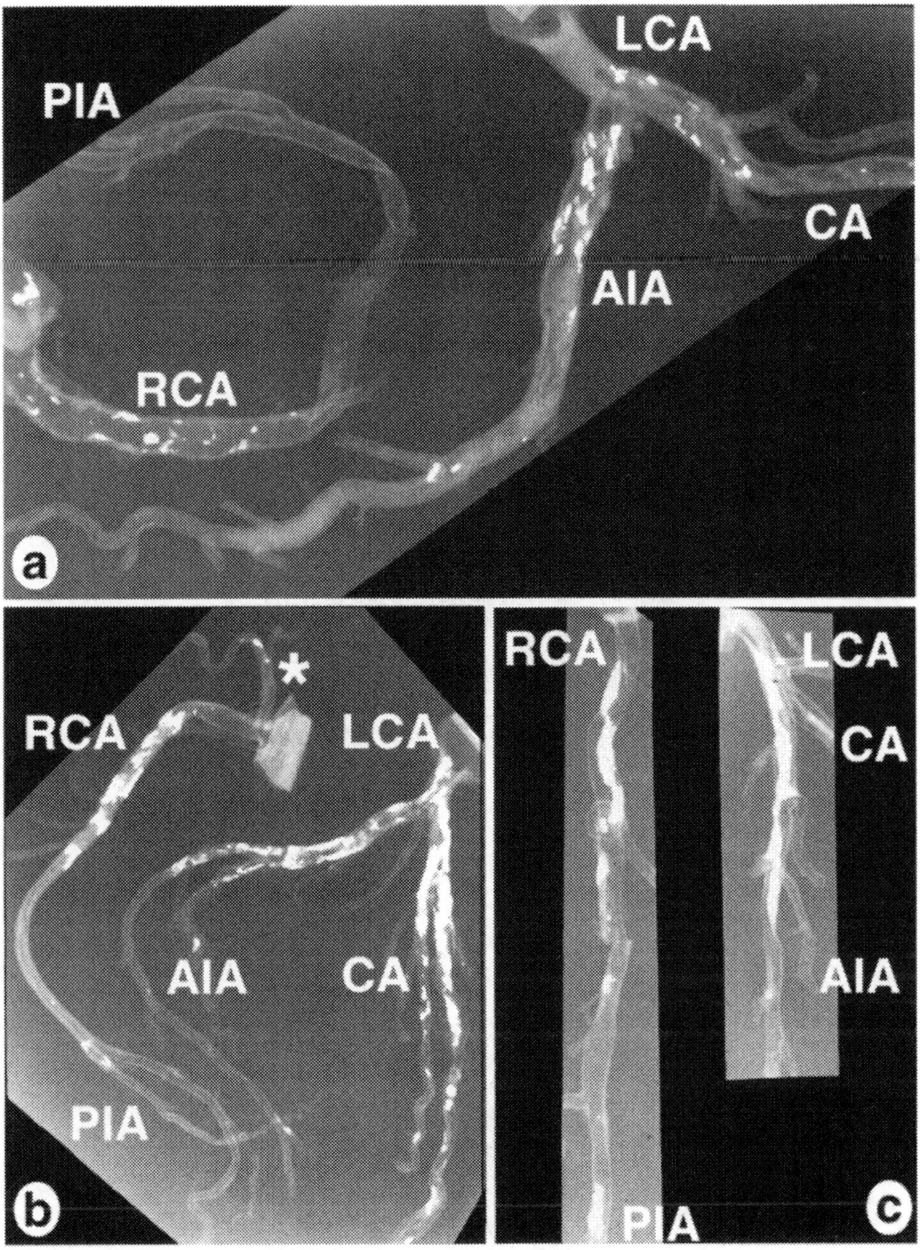

Fig. 26a–c. Calcification deposits in the walls of the coronary arteries (postmortem X-rays); the distribution pattern and quality of calcified deposits in the coronary arteries are similar in every heart specimen. **a** Multiple small deposits in the initial segments of the *AIA* and *CA*, but not in the main stem of the *LCA*; a few small deposits in the *RCA*. Corresponds to a light degree of atherosclerosis. **b** Multiple medium-sized deposits in all three coronary arteries corresponding to a moderate degree of atherosclerosis. Both the first centimeter-long segment of the *RCA* and the origin of the SA node artery are unaffected (*asterisk*). **c** Severe coronary calcification of all parts of the *RCA* and *LCA* forming stripes of 2–3 cm in length. This finding corresponds to a severe degree of atherosclerosis with stenosis of more than 50% of the diameter

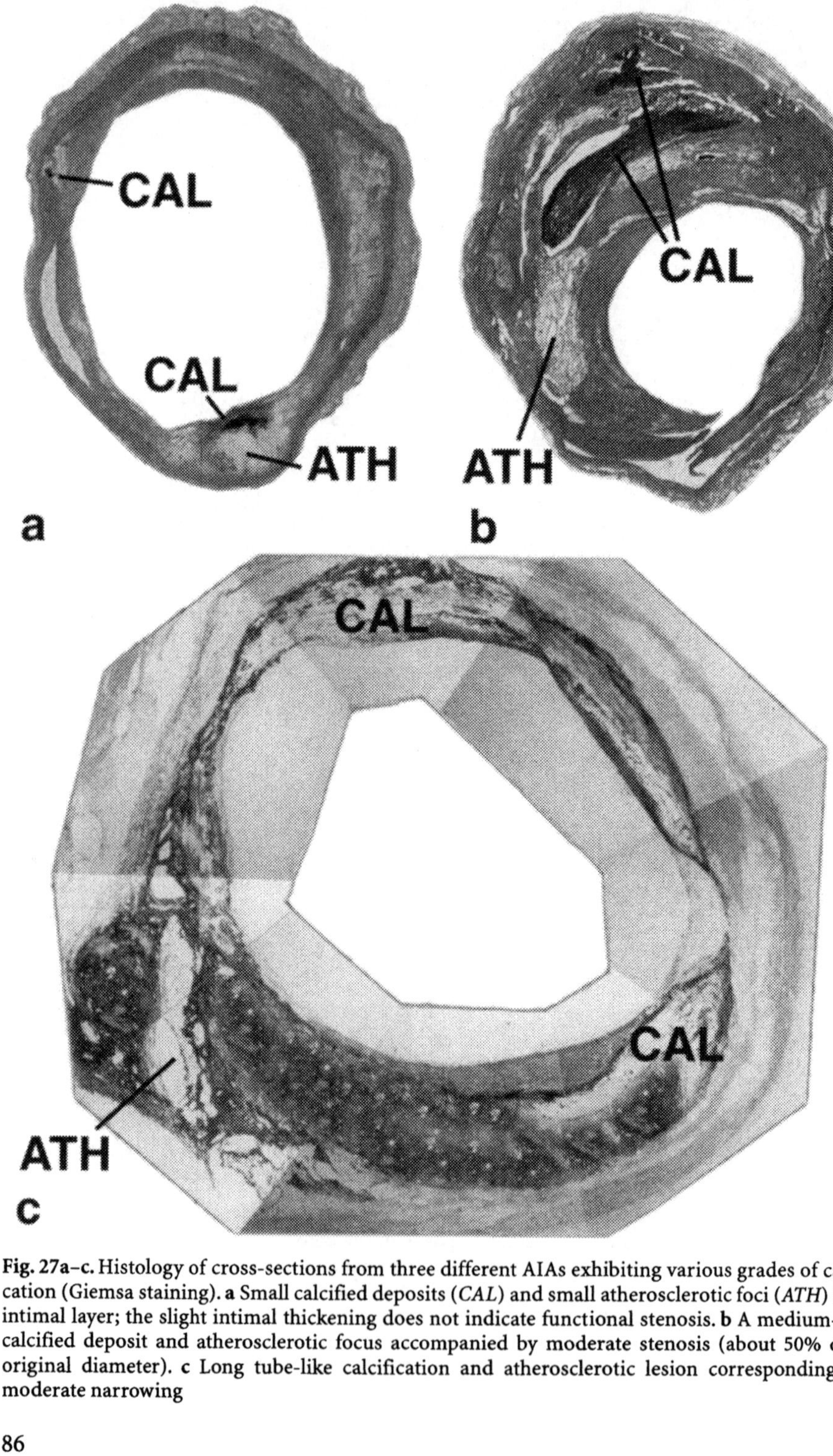

Fig. 27a–c. Histology of cross-sections from three different AIAs exhibiting various grades of calcification (Giemsa staining). **a** Small calcified deposits (*CAL*) and small atherosclerotic foci (*ATH*) in the intimal layer; the slight intimal thickening does not indicate functional stenosis. **b** A medium-sized calcified deposit and atherosclerotic focus accompanied by moderate stenosis (about 50% of the original diameter). **c** Long tube-like calcification and atherosclerotic lesion corresponding to a moderate narrowing

development of atheromatous stenosis of more than 50% of the diameter of an artery in patients over 60 years of age (Blankenhorn 1961; Frink et al. 1970; Noltenius et al. 1976; Masuda et al. 1991; Bormann et al. 1992; Mautner et al. 1994).

In conclusion it can be said that calcification deposits are an almost universal occurrence in the aged; this is not dependent on gender and represents an age-related degenerative process rather than a disease process. Some patients even exhibit a triad of cardiac calcifications involving the aortic cusps, mitral annulus, and coronary arteries, which is designated as "senile calcification syndrome" (Kitzman and Edwards 1990).

4.10.6
Multiple Coronary Aneurysmata

In very rare cases coronary atherosclerosis is combined with local ectasia, dilation and the development of aneurysms of the epimurally distributed arteries (Fig. 25b).

Coronary artery aneurysms have a poor prognosis because of the possibility of rupture with resultant cardiac tamponade, or the development of thrombo-embolic myocardial infarction (Scott 1948; Koischwitz et al. 1982).

4.11
Anatomical Aspects of Surgical Interventions in Coronary Atherosclerosis

4.11.1
Surgical Treatment of Coronary Atherosclerosis

Since pathological lesions of the coronary arteries have been demonstrated with great precision both anatomically (by dissection) and angiographically, bypass surgery using aorto-coronary saphenous vein grafts or thoracia interna anastomoses has enjoyed rapid development and, most importantly, success (Arnulf 1972; Effler and Sheldon 1974; Fallavollita et al. 1994).

Most patients suffering from coronary atherosclerosis exhibit double or triple vessel disease with narrowings in the proximal and/or distal vascular segments: there are moderate, severe, and extreme combinations of obstructive atherosclerosis and other cardiovascular dysfunctions. The combination of acute coronary insufficiency with severe lesions in the first segments of the three main coronary arteries and a corresponding insufficient left ventricular function, together with respiratory distress syndrome, constitutes the worst-case scenario and requires urgent surgical intervention, for instance a triple coronary bypass operation (Conti et al. 1973; Harlan et al. 1983; von der Emde 1991).

4.11.2
Aorto-Coronary Bypass: Sites of Aortic Anastomoses

Saphenous vein grafts providing a triple aorto-coronary bypass are implanted in the anterior wall of the ascending aorta, at three locations from caudal to cephalad (proximal to distal):

1. The aortic anastomosis is the most caudal in the case of the RCA bypass graft, which courses vertically over the right atrium and right part of the coronary sulcus and around the right margin of the heart.
2. The AIA graft has its aortic anastomosis cephalad to the right graft and extends caudally and gradually leftwards, draping over the right ventricular outflow tract or proximal main pulmonary artery before running caudally and towards the AIA, finally reaching the anteroseptal territory.
3. The circumflex saphenous vein graft has the most cephalad aortic anastomosis and courses leftwards and posteriorly to circumflex marginal branches (Aurigemma et al. 1989).

In contrast, left thoracica interna grafts extend caudally and posteriorly from the chest wall towards the territory of the AIA.

4.11.3
Aorto-Coronary Bypass: Sites of Coronary Artery Anastomoses

The typical sites for the surgical implantation of bypass vessels are always located distal to the occlusions or severe stenoses and, if possible, distal to the many calcification deposits of the three coronary arteries.

The surgical treatment of coronary heart disease with aorto-coronary- or thoracica interna (mammaria)-coronary bypass operations can sometimes be avoided due to numerous arterial collateral anastomoses (von der Emde 1991).

5 Concluding Remarks

5.1
Individual Coronary Anatomy

In their daily work, coronary surgeons and cardiologists are constantly confronted with the most variable and anomalous origins and distribution patterns of the coronary arteries. Not only are the positions of the ostia of the coronary arteries important, but also the variations in the angles of arterial take-off, initial courses through the aortic walls, subsequent courses, and the presence of ostial valve-like ridges or membranes that may result in the significant compromise of blood flow. In modern diagnostics the arterial supply to the ventricles and to the atria may be visualized, for example, by selective coronary arteriography. Coronary angiography reveals that no cardiac territory is served by homogeneous vascularization of a well-defined type. Thus, individual coronary anatomy has a great influence on diagnostic procedures, surgical management, and the postoperative progress of a patient who has been subjected to cardiological diagnostics and undergone surgical correction of a congenital vascular or valvular anomaly or acquired vascular or valvular lesion (Effler and Sheldon 1974; Engel et al. 1975).

5.1.1
Coronary Artery Dominance

Hearts which exhibit a particular coronary artery dominance or preponderance are considered to be more susceptible to coronary heart disease (James 1965; Blunk and DiDio 1971). The pattern of vascular distribution influences pathological alterations in the arteries owing to the greater burden assumed by the dominant artery. Therefore, left and right coronary dominance may constitute a significant determinant where prognosis in cases of acquired coronary heart disease is concerned. In a case of LCA dominance the RCA is usually small and often fails to reach the right margin of the heart; thus an acute, proximal occlusion will have disastrous consequences, since the potential for rapid development or reopening of collateral vessels is likely to be diminished (Allwork 1987).

Functionally, the main stem of the LCA is always considered a "dominant" artery. Its atherosclerotic alterations (which are obviously not so frequent) greatly increase the risks in diagnosis and therapy and diminish prognostic expectations (Loop 1981).

In a heart specimen with exceptional right coronary dominance, when the SA node artery and the AV node artery derive from the RCA, proximal occlusion may affect the conduction system and may lead to the disruption of cardiac rhythm and in the worst cases to lethal cardiac arrest (von Lüdinghausen 1975).

With the existence of a pathological condition the division of coronary arteries into the types "dominant right," "dominant left," and "intermediate" becomes imprecise and "blurred" because of the development of innumerable collaterals (Hort and Kalbfleisch 1975).

A few authors criticize the classifications of dominant right and left coronary arteries; they argue that arterial preponderance does not provide clinically significant information for a good systematization of coronary arterial irrigation (Reig et al. 1993b).

5.1.2
Right Ventricular Branches of the AIA

In most cases the right ventricular branches of the AIA are small, tenuous, and lack clinical significance. However, Cabin et al. (1987) found large right ventricular branches of the AIA in 13% of the cases they studied, which had caused large areas of right ventricular myocardial infarction.

5.1.3
Atrial Branches

The distribution pattern of atrial branches (except for the branches involved in the conduction system) has almost been neglected in modern cardiological literature; atrial infarction is not so rare, as it occurred in 17% of their cases with a unilateral predominance of auricular infarction (27 cases in the right atrium and 4 in the left). By contrast, Blumgart et al. (1940) explained the comparative rarity of infarction in the right heart chambers – in spite of the relatively large number of cases exhibiting occlusion of the RCA – as due to the thin walls, which enable the atrial and ventricular myocardium to obtain a considerable supply of blood from its own cavity through the great number and variety of extracardiac anastomoses and collaterals. These were found angiographically in 48% of the cases of coronary artery disease examined by Björk (1966) and Dupont and Riquet (1991); in 12%, the anastomoses were over 2 mm in diameter.

It should be remembered that the RCA supplies not only the right atrial and right ventricular myocardium but also (in 60% of cases generally) the sinus node and (in 90%) the AV node. The important point is that the sudden development of heart block in a patient with coronary disease virtually always indicates that the RCA is occluded or markedly narrowed.

Recent developments and advances in experiments and human cardiac surgery – for instance surgery of the conductive elements and in cases of Wolff-Parkinson-White-Syndrome – have led to an increase in the amount of attention paid to the atrial branches in the crux cordis area (DiDio and Wakefield 1972; Cox and Ferguson 1989): the AV node artery is threatened when an accessory atrioventricular conduction route, notably the bundle of Kent in their posterior paraseptal localization, has to be cut (Sow et al. 1996).

The proximity of the AV-node artery to the septal cusps of the tricuspid and mitral valves makes these structures liable to damage if they are included in the sutures of prostheses during mitral and tricuspid valve replacement (Vlodaver et al. 1976; Arid et al. 2000).

5.1.4
Unusual Short Stem of the LCA

The clinical significance of a short LCA is that it may complicate perfusion to the left coronary system during operative procedures such as aortic valve replacement. Despite apparently optimal placement, the cannula may perfuse either the AIA or the CA but not both, resulting in myocardial ischemia with ventricular arrhythmias or myocardial infarction or both (Vlodaver et al 1976).

The length of the main LCA was found by Gazetopoulos et al. (1976) to be significantly shorter in patients with coronary atherosclerosis than in patients without angiographical evidence of coronary artery disease. A short left main LCA means that there is an increased risk of the development of severe coronary atherosclerosis. This finding is considered to constitute a congenital factor predisposing the development of coronary heart disease.

An anomalous CA, which originates ectopically from the proximal RCA and surrounds the aortic bulb before following a course in the left coronary sulcus, may be occluded during aortic valve replacement when the valve prosthesis is implanted and fixed by numerous stitches.

5.1.5
Normally or Anomalously Coursing Coronary Arteries in Valve Surgery

Normally or anomalous coursing coronary arteries adjacent to the fibrous valve rings – for instance the normal CA in the left coronary sulcus or the anomalous CA with right coronary origin – have to be taken into consideration during surgery. Because of their proximity to the base of the mitral or aortic valve, these vessels are endangered during valve prosthesis implantation (Black et al. 1973; Vlodaver et al. 1976; Engel et al. 1975; Engel and Dalichau 1976). Others are hemodynamically significant and alter myocardial perfusion. The so-called primary anomalies are, for instance, a single coronary artery (the other originating from the "right" heart), coronary artery fistula, and the origin of the LCA from the right aortic sinus, with subsequent passage of the vessel between the aorta and the right ventricular infundibulum (Levin et al. 1978).

5.1.6
Single Coronary Artery

A single coronary artery and an origin of both coronary arteries from the same aortic sinus have been traditionally regarded as having little clinical significance and as compatible with a long and active life. There is a left and a right type of this anomaly, connected with either the left or right aortic sinus. However, recognition and angiographic demonstration of these anomalies assume the highest priority in a patient undergoing evaluation for direct coronary artery surgery or prosthetic valve replacement . For instance, there is the question of whether each aberrant RCA of this kind – when replacement of the mitral valve is indicated – may be susceptible to ligature or to compression by prosthetic rings, leading to infarction in the patient (Roberts and Morrow 1969; Black et al. 1973; Engel et al. 1975; Becker 1995; Karadede et al. 2001).

The condition of a single RCA deriving from the aorta and of an aberrant LCA deriving from the pulmonary trunk usually causes early fatality; this is designated as Bland-White-Garland syndrome (Bookstein 1964).

5.1.7
Multiple Ostia in the Aortic Sinuses

The occurrence of multiple ostia in the aortic sinuses is quite frequent and probably has minimal clinical relevance, except that attempts to carry out selective coronary injections may lead to the erroneous conclusion that an artery is missing (Becker 1995).

5.1.8
Ectopic "High" Origin

Ectopic (superior) origin, "high take-off" and transposition of the right or left coronary ostia occasionally cause diminished myocardial perfusion, myocardial ischemia and even sudden death. The possible reason for the latter could be that the extremely acute angle at which the ectopic artery leaves the aorta may cause a flap-like mechanism at the arterial ostium (Mahowald et al. 1986).

5.1.9
Ectopic "Deep" Origin

Ectopic deep (inferior) origin or "low take-off" of a coronary artery from the floor of the aortic sinus may cause an ostial stenosis and even blockage following aortic valvular replacement. During and after surgery the newly-implanted prosthesis may obstruct the ostium of one or both coronary arteries and may cause intra- or postoperative complications or even sudden death (Black et al. 1973; Vlodaver et al. 1976).

5.1.10
Myocardial Bridge and Intramyocardial (Intramural) Course

A short myocardial bridge of the AIA (i.e., short intramyocardial course) is a frequent occurrence; it may be identified angiographically because it exhibits a typical U-turn. A superficial bridge does not seem to constrict the related artery during systole, but a deeply submerged myocardial bridge may affect (twist) the vessel during systole. This can create a phenomenon of "milking," whereby the artery is presumably constricted during systole and widens during diastole. Functionally no stenosis is present, and myocardial bridging should not be confused with a long segment of stenosis (Greenberg et al. 1989). The "milking" phenomenon may result in ischemia (McAlpine 1975; Allwork 1987; Ferreira et al. 1991). The degree of compression varies from negligible to severe, as expressed by a spectrum of pathology ranging from angina, through tachyarrhythmias to myocardial infarction. There seems to be an association between sudden death and vigorous exercise in otherwise healthy young people and this anatomical variant (Frauenfelder and Schneider 1982; Allwork 1987).

Proximal to the bridge, the intima of the artery is 30% thicker than it is in the subjacent segment in the absence of intraluminal disease (Polacek 1961). When disease is present the proximal segments are more severely affected than those subjacent and distal to the bridge (Stolte et al. 1977).

Within its intramural course or courses a coronary artery is practically free from intimal thickening or disease and shows few or no signs of atherosclerosis such as intimafibrosis, atheromas, and calcification deposits (Fulton 1965; von Lüdinghausen and Valeske 1978). It was observed that "immunity" from atherosclerotic change was only conferred on those arteries which were covered by a layer of substantial thickness, about 1 mm and more, and that the protective effects of muscle layers of lesser thickness were variable. However, some authors (for example, Edwards et al. 1956) assume that the intramyocardial portions of the coronary arteries are not protected, since atherosclerotic processes could occur there just as in extramural portions, and also in other entirely extramural coronary arteries.

5.1.11 Interventricular and Coronary Sulci Without an Epimural Artery

The existence of a long intramyocardial and intracavitary course of a main coronary artery indicates that the coronary and interventricular sulci contain no epimural artery.

During a bypass operation and grafting of a bypass vessel a deep intramural course of a coronary artery and artery-free sulcus will be potentially disconcerting to the coronary surgeon who is searching for an arterial vessel on the surface of the recipient heart.

An AIA coursing deeply in the subendocardial layers will not only be disconcerting to the coronary surgeon, but also enhances the risk of sudden ventricular bleeding (McAlpine 1975).

In younger patients the existence of myocardial bridges may be paralleled by/appear to relate to the evident obstruction of the segment concerned during systole. During diastole, the obstruction may disappear (Vlodaver et al. 1976).

5.1.12 Intra-, Inter-, and Noncoronary Extracardiac Anastomoses and Collaterals

Tenuous, preexisting anastomoses between the branches of the coronary arteries in every heart may widen and develop collaterals in cases exhibiting severe obstructive atherosclerosis. Some collaterals are newly formed as a response to ischemia (Debiec

1967; Allwork 1987; Dupont and Riquet 1991). Anastomoses and collateral circulatory channels are frequently found in the IV septum and in the atrial walls (Petelenz 1965b; Reig-Vilallonga et al. 1987; Reig et al. 2000).

Because of their multiple anastomoses to other coronary branches, the arteries with ectopic derivation from the right aortic sinus (right conus branch/artery) and the arterial connections with a bronchial artery may achieve a significant role as bypass vessels in cases of severe coronary atherosclerosis (Petelenz 1965b; Rathor et al. 1973; Esperenca Pina 1974; McAlpine 1975; von Lüdinghausen and Ohmachi 2001).

Although the potential importance of noncoronary extracardiac circulation is demonstrated in patients who, despite total occlusion of all three major coronary arteries, maintain good ventricular function (Leary and Wearn 1930; Brazier et al. 1975), the exact role of collateral circulation is not known. Some investigators have found no difference in overall left ventricular function, with or without the presence of collaterals (Carroll et al. 1974). The development of collateral circulation is occasionally, but not always, associated with the restoration of left ventricular function. Angiographically, there is a positive correlation between the quality of collateral filling and preservation of the left ventricle wall motion in the area supplied (Hecht et al. 1975). Therefore, it may only have a beneficial effect on myocardial supply (Helfant et al. 1971; Smith et al. 1972; Carroll et al. 1974).

The widely spread subendocardial plexuses which are connected with the smallest cardiac vessels (formerly Thebesian vessels) appear to play an important role as collateral channels in patients with coronary heart disease (Estes et al. 1966a).

Numerous extracardiac vessels serving the myocardium have been found in the newborn human (Halpern 1954). In cases with chronic fibrous pericarditis multiple collateral vessels within adhesions between the peri- and epicardium may nourish larger areas of myocardium and in this way replace the blood flow of the coronary arteries (Beck 1935; Schildt et al. 1943; Harken et al. 1955; Bloor 1974).

5.1.13
Coronary Artery Calcification

Coronary calcification deposits are numerous in angiographically treated patients between 35 and 70 years of age and anatomically investigated subjects between 65 and 91 years of age at death (Montenegro and Eggen 1968; von Lüdinghausen and Valeske 1979).

Current opinion accords calcification an ancillary role in the genesis of atherosclerosis of coronary arteries. The increase of calcification with advancing age is usually indicative of multiple vessel disease; this is especially the case if there is a marked presence. In proportion to its degree of severity, the likely association/correlation of calcification of the LCA with proximal stenosis of the AIA increases. Severe calcification of the RCA is not a common finding but its presence is consistent with proximal stenosis of this vessel.

Thus, many clinicians state quite clearly that "the predominant cause of coronary artery calcification is atherosclerosis" (Blankenhorn 1961; Frink et al. 1970; Bierner et al. 1978; Fallavollita et al. 1994).

Davies (1994) and Margolis et al. (1980) found the 5-year survival rate of patients with severe coronary calcification, when compared to that of patients in whom calcification could not be detected, to be almost worse, independent of the extent of disease exhibited by the two groups.

The lack of vascular lesions in the first 1 and 2 cm of the coronary arteries in the group exhibiting mild and moderate calcific deposits in our material is conspicuous. This indicates that the many SA node arteries originating in a very proximal position (close to the aortic ostium) are not affected by calcified atherosclerotic lesions.

Where the frequent distally situated origins of the SA node and AV node branches are concerned, calcifications and narrowing in plaques near to the aortic ostium jeopardize the blood supply to the SA node and AV node. Thus, arterial ramification patterns and the location of plaques, calcifications, and stenoses often correspond to, and depend on, each other (von Lüdinghausen and Valeske 1979; Masuda et al. 1991; Mautner et al. 1994).

The AIA is the vessel which most frequently develops calcification; less frequently the calcifications are seen in the right coronary and in the left circumflex arteries.

Preoperative detection of a severe grade of calcification (for instance by ultrafast CT scan) is quite essential in coronary bypass surgery, because no suture can be fixed to a rigid calcified arterial wall.

Meticulous consideration of minute anatomical details helps to minimize the risk of accidental ablation of important arteries and parts of the cardiac conduction system during intracardiac repair, valve replacement, bypass implantation, and treatment of WPW syndrome.

It is of great advantage anatomically that the right and left SA node arteries arise from the first 1-cm-long segment of either the RCA or CA because atherosclerotic lesions frequently develop in the second and third cm segments distal to the aortic ostium. However, when the SA node arteries originate "late," far distal to the calcifying stenosis (as is always the case with the AV node artery) then there is a danger of circulatory turbulence in the conduction system.

In cases with severe atherosclerosis of the coronary arteries, an extensive extracardiac arterial supply of the atrial myocardium (including the SA node area) and adjacent ventricular myocardium may develop to replace original coronary function (Debiec 1967; Moberg 1967a,b).

5.1.14 Evaluation and Interpretation

Although it has become popular to assume a relation between a given epimural coronary artery (an "infarct artery"), and its myocardial supply (the "myocardium at risk"), consideration has to be given to the fact that the amount of myocardium supplied by one vessel is variable and is affected by anastomosing networks and collateral vessels, congenital variations, and other factors (Kitzman and Edwards 1990).

Some anatomical findings seem to have an influence where the development of coronary atherosclerosis is concerned: the shortness of the main LCA and some coronary anomalies may amplify the severity of the disease (Gazetopoulos et al. 1976).

With regard to obstructive coronary heart disease with calcifications, occlusion and myocardial infarction there has been some tendency to consider the AIA as the "artery of sudden death," though subsequent observations have not indicated a sound basis upon which this intriguing designation could be substantiated. There are now some arguments for the assumption that the designation may apply to the RCA (Vlodaver et al. 1976). Since the RCA supplies the SA node in 60% and the AV node in 90% of human hearts, many instances of heart block with or without escape ventricular rhythms are due to this occlusion (James 1965; Isner and Roberts 1978).

Severe complete obstruction of the main stem of the LCA can cause sudden death from irreversible pump failure. Therefore this artery has also been designated "the artery of sudden death" (Gotsman et al. 1973).

The clinical significance of severe and extensive calcification deposits in the coronary artery walls is that (1) they correlate well with the degree of coronary artery stenosis (Mautner et al. 1994), (2) they indicate the extent of atherosclerosis and are crucial for prognosis in patients at risk (Mautner et al.1994; Baron 1994; Knez 2000), and (3) they determine surgical procedure at the reconstruction site. The distal end of a bypass vessel must be implanted distally to the narrowings and calcification deposits; otherwise, there is a high risk of break-up of the graft and of massive pericardial hemorrhage and myocardial ischemia (Effler et al. 1967; von Lüdinghausen and Valeske 1978).

The experience of the majority of cardiologists and clinicians is directly the opposite of that of a few authors who emphasize that coronary artery calcification observed in an unselected group of individuals is not associated with evidence of impaired coronary function (Pyke and Symons 1951).

6 Summary

In view of the fact that, in cardiology and coronary surgery, peculiarities and anomalies of the coronary arteries pose a great potential risk for myocardial ischemia resulting in arrhythmia, angina, infarction, and sudden death, a survey of the morphology of the coronary arteries in cadaver specimens (n=200) and corrosion casts (n=30) was undertaken. The results are not divided into those obtained from subjects with or without coronary heart disease. The randomly selected material from two anatomical institutes, of the Universities of Munich and Würzburg, was considered representative of an average group of septuagenarians and octogenarians in the German population. It is the patient of this age that the young medical students of today (the future physicians) will be confronted with and called upon to treat.

The study begins with a review of the nomenclature and a consideration of unofficial (but often clinically used) terms and synonyms for coronary arteries and their branches, followed by a description of modern diagnostic techniques based on the presence of normal and anomalous cardiac vessels. It continues with a demonstration of „normal“ macro- and microanatomy, a description of the most frequent ramification pattern and as well of the many peculiarities of the coronary arteries. These are briefly enumerated here: ectopic coronary ostia and aberrant and accessory arteries; angles of origin and division; three-dimensional orientation of the coronary stems; intramural, intraluminal, and intracavitary courses; avascular coronary and interventricular sulci; extracardiac noncoronary myocardial supply; atrial branches, sinuatrial node artery, atrioventricular node artery, principles of septal arterial supply, variations of the right and left superior septal arteries; endomural, intracoronary, intercoronary, and noncoronary anastomoses and collaterals; measurement and evaluation of age-related and atheromatous coronary calcifications; multiple coronary aneurysmata.

References

Adachi B (1928) Das Arteriensystem der Japaner, Bd 1. Kaiserlich-Japanische Universität, Kyoto, pp 17–22

Alfidi RJ, Masaryk TJ, Haacke EM, Lenz GW, Ross JS, Modic MT, Nelson AD, LiPuma JP, Cohen AM (1987) MR angiography of peripheral, carotid and coronary arteries. Amer J Roentgenol 149: 1097–1109

Ahmed S, El-Rakhawy, Abdalla A, Harrison RG et al (1972) A new conception of coronary artery preponderance. Acta Anat (Basel) 83: 87–94

Akins EW, Hill JA, Fitzsimmons JR, Pepine CJ, Williams CM (1985) Importance of imaging plane for magnetic resonance imaging of the normal left ventricle. Amer J Cardiol 56: 366–372

Alexander R, Griffith G (1956) Anomalies of the coronary arteries and their clinical significance. Circulation 14: 800–805

Allen GL, Snider TH, Houston FS (1966) Myocardial infarction with a single coronary artery. Arch Intern Med 117:261–264

Allwork SP (1979) A spectrum of normal coronary artery distribution in man. Anat Clin 1: 311–319

Allwork SP (1987) The applied anatomy of the arterial blood supply to the heart in man. J Anat 153: 1–16

Anderson CM, Edelman RR, Turski PA (1993) (eds) Clinical magnetic resonance angiography. Raven Press, New York, pp 419–425

Anderson KR, Ho SY, Anderson RH (1979) Location and vascular supply of the sinus node in the human heart. British Heart J 41: 28–32

Anderson RH, Becker AE (1982) Anatomie des Herzens. Thieme, Stuttgart, pp 129–131

Angelini P (1989) Normal and anomalous coronary arteries: definitions and classification. Amer Heart J 117: 418–434

Arid JM, Armstrong O, Rogez JM, Robert R, Lardoux MC, Leborgne J (2000) Arterial vascularisation of the atrio-ventricular node. Surg Radiol Anat 22: 93–96

Arnett EN, Isner JM, Redwood DR, Kent KM, Baker WP, Ackerstein H, Roberts WC (1979) Coronary artery narrowing in coronary heart disease: comparison of cineangiographic and necropsy findings. Ann Int Med 91: 350–356

Arnulf G (1972) The evolution of the surgical treatment of coronary arteriosclerosis. J Cardiovas Surg 13: 373–381

Atkinson DJ, Burstein D, Edelman RR (1990) First-pass cardiac perfusion: evaluation with ultrafast MR imaging. Radiology 174: 757–762

Aurigemma GP, Reichek N, Axel L, Schiebler M, Harris C, Kressel HY (1989) Non-invasive determination of coronary artery by-pass graft patency by cine magnetic resonance imaging. Circulation 80: 1585–1602

Bader G (1963) Beitrag zur Systematik und Häufigkeit der Anomalien der Coronararterien des Menschen. Virchows Arch Path Anat 337: 88–96

Bakst A, Lewis BS, Mitha AS, Gotsman MS (1974) Isolated obstruction of the right coronary artery. Chest 65: 18–24

Banchi A (1904) Morfologia delle arteriae coronariae cordis. Arch Ital Anat Embriol 3: 87–164

Bankl H (1977) Congenital malformations of the heart and great vessels. Urban and Schwarzenberg, Baltimore, Munich, pp 213–220

Baptista CAC, DiDio LJA, Davis JT, Teofilovski-Parapid G (1988) The cardiac apex and its superficial blood supply. Surg Radiol Anat 10: 151–160
Baptista CAC, DiDio LJA (1992) The relationship between the directions of myocardial bridges and of the branches of the coronary arteries in the human heart. Surg Radiol Anat 14: 137–140
Bargmann W (1963) Innervation des Herzens. In Bargmann W, Doerr W (eds) Das Herz des Menschen, Bd I. Thieme Stuttgart, pp 136–137
Barmeyer J (1968) Postmortale Koronarangiographie und Perfusion normaler und pathologisch veränderter Herzen; Messung der Durchflusskapazität interkoronarer Anastomosen. Beitr Path Anat 137: 373–390
Baroldi G, Mantero O, Scomazzoni G (1956) The collaterals of the coronary arteries in normal and pathologic hearts. Circulation Res 4: 223–229
Baroldi G, Scomazzoni G (eds.) (1965) Coronary circulation in the normal heart. In: Coronary circulation in the normal and pathologic heart. Washington D.C.: Armed Institute of Pathology, Office of the Surgeon General, Department of the Army, pp 5–73
Baroldi G (1974) Distribution of arterio-atherosclerotic obstructive lesions in coronary arteries.Vasc Surg 53: 54–59
Baron MG (1994) Significance of coronary artery calcification. Radiology 192: 613–614
Barry A, Patten BM (1953) The structure of the adult heart. In Gould SE (ed) Pathology of the heart. CC Thomas, Springfield Ill, pp 87–125
Beck CS (1935) The development of a new blood supply to the heart by operation. Ann Surg 102: 801–813
Becker AE (1995) Congenital coronary arterial anomalies of clinical relevance. Coronary Artery Disease 6: 187–193
Belman S, Frank HA (1958) Intercoronary collaterals in normal hearts. J Thoracic Surg 36: 584–589
Benson PA, Lack AR (1968) Anomalous aortic origin of left coronary artery. Arch Path 86: 214–216
Benson PA (1970) Anomalous aortic origin of coronary artery with sudden death: case report and review. Amer Heart J 79: 254–257
Bertho E, Gagnon G (1964) A comparative study in three dimension of the blood supply of the normal interventricular septum in human, canine, bovine, porcine, ovine, and equine heart. Dis Chest 46: 251–262
Bezerra AJC, Prates JC, DiDio LJA (1987) Incidence and clinical significance of bridges of myocardium over the coronary arteries and their branches. Surg Radiol Anat 9: 273–280
Bezerra AJC, DiDio LJA, Piva Jr L (1989) Myocardial bridge over the right coronary artery in man. Surg Rad Anat 11: 271–273
Bierner M, Fleck E, Dirschinger J, Hall D, Klein U, Rudolph W (1978) Relationship between coronary artery calcification and significant stenosis. III. Symposium on coronary heart disease, Sept 8–11, 1978, Frankfurt/M. (abstract)
Björk L (1966) Angiographic demonstration of extracardial anastomoses to the coronary arteries. Radiology 87: 274–277
Björk L (1969) Angiographic demonstration of collaterals to the coronary arteries in patients with angina pectoris. Acta Radiol Diagn 8: 305–309
Black LL, McComb RJ, Silver MD (1973) Vascular injury following heart valve replacement. Ann Thorac Surg 16: 19–29
Blankenhorn DH (1961) Coronary arterial calcification. Amer J Med Sciences 242: 41–50
Bloor CM, Liebow AA (1965) Coronary collateral circulation. Amer J Cardiol 16: 238–252
Bloor CM (1974) Functional significance of the coronary collateral circulation. Amer J Pathol 76: 562–586
Blumenthal HT, Lansing AI, Wheeler PA (1944) Calcification of the media of the human aorta and its relation to intimal arteriosclerosis, ageing, and disease. Amer J Pathol 20: 665–687
Blumgart HL, Schlesinger MJ, Davis D (1940) Studies on the relation of the clinical manifestations of angina pectoris, coronary thrombosis, and myocardial infarction to the pathologic findings. Amer Heart J 19: 1–91
Blunk JN, DiDio LJA (1971) Types of coronary circulation in human hearts. Ohio State Med J 67: 596–607

Bogers AJJC, Gittenberger de Groot AC, Poelmann RE, Peault PM, Huysmans HA (1989) Development of the origin of the coronary arteries, a matter of ingrowth or outgrowth ? Anat Embryol 180: 437–441
Bookstein JJ (1964) Aberrant left coronary artery. Amer J Roentgenol 91: 515–528
Bormann JL, Stanford W, Stenberg RG, Winniford MD, Berbaum KS, Talman CL, Galvin JR (1992) Ultrafast computed tomographic detection of coronary artery calcification as indicator for stenosis. Amer J Cardiac Imaging 6: 191–196
Brazier J, Hottenrot C, Buckberg G (1975) Non-coronary collateral myocardial blood flow. Ann Thorac Surg 19: 426–435
Bream PR, Souza AS, Elliott LP, Soto B, Curry GC (1979) Right superior septal perforator artery: its angiographic description and clinical significance. Amer J Roentgenol 133: 67–73
Brundage B (1994) Computed tomography of the heart. In Schlant RC, Alexander RW (eds) Hurst`s The heart, 8th ed. McGraw-Hill, New York, pp 2325–2338
De Brux JL, Pernes JM, Grenier P (1987) Magnetic resonance imaging of the heart compared with anatomic and ultrasonographic data. Surg Radiol Anat 9: 303–314
Byrd BF, Finkbeiner W, Bouchard A, Silverman NH, Schiller NB (1989) Accuracy and reproducibility of clinically acquired two-dimensional echocardiographic mass measurements. Amer Heart J 118: 133–137
Cabin HS, Clubb KS, Wackers FJT, Zaret TL (1987) Right ventricular myocardial infarction with anterior wall left ventricular infarction: an autopsy study. Amer Heart J 113: 16–23
Cabrol C, Christides C (1975) Vascularisation des noeuds sino-auriculaire et atrioventriculaire. Arch Mal Coeur 12: 1237–1242
Campbell JS (1929) Stereoscopic radiography of the coronary system. Quarterly J Med 22: 247–267
Carroll RJ, Verani MS, Falsetti HL (1974) The effect of collateral circulation on segmental left ventricular contraction. Circulation 50: 709–713
Castorina S (1957) (cited in Allwork 1987)
Chaitman BR, Lesperance J, Saltiel J, Bourassa MG (1976) Clinical, angiographic and hemodynamic findings in patients with anomalous origin of the coronary arteries. Circulation 53: 122–131
Cheitlin MD, de Castro CM, McAllister HA (1974) Sudden death as a complication of anomalous left coronary origin from the anterior sinus of Valsalva. Circulation 50: 780–787
Cheitlin MD (1989) Coronary arterial anomalies: clinical and angiographic aspects. In: Virmani R, Forman MB (eds) Non-atherosclerotic ischemic heart disease. Raven Press, New York, pp 125–151
Cho ZH, Mun CW, Friedenberg RM (1991) NMR angiography of coronary vessels with 2-D planar image scanning. Magn Res Med 20: 134–143
Christ B (1990) Grundlagen der embryonalen Gefäßbildung. In Hinrichsen (ed) Human-embryologie, Springer Berlin, Heidelberg, New York, pp 247–248
Clarke JA (1965) An X-ray microscopic study of the pattern and distribution of capillary beds in the atria of the human heart. Z Anat Entw.Gesch 124: 471–477
Clausen M, Civelek AC, Bice AN, Petronis J, Koller D, Loncaric S, Weiss JL, Wagner HN (1988) Short-axis circumferential profiles of the heart in healthy subjects: comparison of Tl-201 SPECT and 2-dimensional echocardiography. Radiology 168: 723–726
Click RL, Holmes DR, Vlietstra RE, Kosinski AS, Kronmal RA (1989) Anomalous coronary arteries: location, degree of atherosclerosis and effect on survival. J Amer Coll Cardiol 13: 531–537
Cohen MV (1985) Coronary collaterals, clinical & experimental observations. Futura Publishing Company, Mount Kisco, New York, pp 1–91
Conte G, Pellegrini A (1984) On the development of the coronary arteries in human embryos, stages 14–19. Acta Embryol 169: 209–218
Conti CR, Brawley RK, Griffith LSC, Pitt B, Humphries JO, Gott VL, Ross RS (1973) Unstable angina pectoris: morbidity and mortality in 57 consecutive patients evaluated angiographi-cally. Amer J Cardiol 32: 745–750
Conti CR (1977) Coronary angiography. Circulation 55: 227–237
Corbett JR, Jansen DE, Willerson JT (1987) Radionuclide ventriculography: I. technical aspects. Amer J Physiol Imaging 2: 33–43
Cornel JH, Ten Cate FJ, Serruys PW (1992) Myocardial contrast echocardiography can depict Thebesian vein outflow in humans. Amer Heart J 123: 1373–1374

Coussement P, de Geest H (1994) Multiple coronary artery – left ventricular communications: an unusual prominent Thebesian system. Acta Cardiol 49: 165–173
Cox JL, Ferguson Jr TB (1989) Surgery for the Wolff-Parkinson-White Syndrome: the endocardial approach. Sem Thoracic Cardiovasc Surg 1: 34–46
Crainicianu A (1922) Anatomische Studien über die Coronararterien und experimentelle Untersuchungen über ihre Durchgängigkeit. Virchows Arch Path Anat 238: 1–75
Crochet D, Lavenant F, Krause D, Wanlin D, Bizais Y, Delumeau, Petitier H (1981) Atheromatous coronary artery disease: value of a study of calcification during assessment . J Radiol 62: 221–228
Crooks LE, Barker B, Chang H, Feinberg D, Hoenninger JC, Watts JC, Arakawa M, Kaufman L, Sheldon PE, Botvinick E, Higgins CB (1984) Magnetic resonance imaging strategies for heart studies. Radiology 153: 459–465
Dauber W (1998) In Feneis H. Anatomisches Bildwörterbuch, 8. Aufl., Thieme, Stuttgart. p 191
Davies JS, Lie JT (1977) Anomalous origin of a single coronary artery from the innominate artery. Angiology 28: 775–778
Davies MJ (1971) Pathology of conducting tissue of the heart. Butterworth, London 1971, pp 1–28
Davies MJ (1994) The pathology of coronary atherosclerosis. In Schlant RC, Alexander RW (eds) Hurst´s The heart, 8th ed. McGraw-Hill, pp 1009–1020
Day RW, Laks H, Drinkwater DC (1992) The influence of coronary anatomy on the arterial switch operation in neonates. J Thorac Cardiovasc Surg 104: 706–712
Dbaly J, Ostadal B, Rychter Z (1968) Development of the coronary arteries in rat embryos. Acta Anat 71: 209–222
Debiec B (1967) Anastomoses of coronary arteries with ramifications of bronchial arteries and the extracoronary vascularization of the Keith-Flack node in the case of children. Aggiornamento Pediatrico 18: 237–249
Deedwania PC, Carbajhal EV (1991) Silent myocardial ischemia: a clinical perspective. Arch Intern Med 151: 2373–2382
DiDio LJA (1967) The atrio-ventricular branches of the human coronary arteries. J Morphol 123: 397–404
DiDio LJA, Wakefield TW (1972) Origin, classification, nomenclature, and incidence of the atrial arteries in normal human hearts with special reference to their clinical importance. Acta Cardiol 27: 565–595
DiDio LJA, Wakefield TW (1975) Coronary arterial predominance or balance on the surface of the human cardiac ventricles. Anat Anz 137: 147–158
DiDio LJA, Rodrigues H (1983) Cardiac segments in the human heart. Anat Clin 5: 115–124
DiDio LJA (1989) Anatomico-surgical segmentation as a principle of construction of the human body and its clinical applications. Anat Anz (Suppl.) 164: 737–743
DiDio LJA, Lopes AC, Caetano AC, Prates JC (1995) Variations of the origin of the artery of the sinoatrial node in normal human hearts. Surg Radiol Anat 17: 19–26
Dollar AL, Roberts WC (1989) Retroaortic epicardial course of the left circumflex coronary artery and anteroaortic intramyocardial (ventricular septum) course of the left anterior descending artery. Amer J Cardiol 64: 828–829
Donaldson RM, Thornton A, Raphael MJ, Sturridge MF, Emanuel RW (1979) Anomalous origin of the left anterior descending artery from the pulmonary trunk. Europ J Cardiol 10: 295–300
Donaldson RM, Raphael MJ (1982) Missing coronary artery: review of technical problems in coronary arteriography resulting from anatomical variants. Br Heart J 47: 62–70
Doty DB; Chandramouli B; Schieken RE; Lauer RM; Ehrenhaft JL (1976) Anomalous origin of the left coronary artery from the right pulmonary artery. J Thorac Cardiovasc Surg 71: 787–791
Douglas PS, Fiolkoski J, Berko B, Reichek N (1988) Echocardiographic visualization of coronary artery anatomy in the adult. J Amer Coll Cardiol 11: 565–571
Duckworth F, Mukharji J, Vetrovec WV (1987) Diffuse coronary artery to left ventricular communications. Cathet Cardiovasc Diagn 13: 133–137
Dumoulin CL, Souza SP, Darrow RD, Adams WJ (1991) A method of coronary MR angiography. J Comput Assist Tomogr 15: 705–710
Dupont P, Riquet M (1991) The bronchial arteries. A review of their anatomy and their anastomoses with the coronary arteries. Surg Radiol Anat 13: 69–71

Dusek J, Ostadal B, Duskova M (1975) Postnatal persistence of spongy myocardium with embryonic blood supply. Arch Pathol 99: 312–317
Edwards JC, Burnsides C, Swarm RL, Lansing AI (1956) Arteriosclerosis in the intramural and extramural portions in the human heart. Circulation 13: 235–241
Edwards JE (1958) Anomalous coronary arteries with special reference to arteriovenous-like communications. Circulation 17: 1001–1006
Edwards JE (1964) Direction of blood flow in coronary arteries arising from the pulmonary trunk. Circulation 29: 163–166
Edwards JE (1968) Congenital malformations of the heart and great vessels. In Gould SE (ed) Pathology of the heart, Charles C. Thomas, Springfield Ill, pp 426–444
Effler DB, Groves LK, Suarez EL, Favaloro RG (1967) Direct coronary artery surgery with endarterotomy and patch-graft reconstruction. J Thorac Cardiovasc Surg 53: 93–101
Effler DB, Sheldon WC (1974) Myocardial revascularization. In Gibbon JH, Sabiston DC, Spencer FC (eds) Surgery of the Chest. WB Saunders Philadelphia, London, Toronto, pp 866–886
Eie H, Hillestad L (1971) Arterio-venous fistulae of the coronary arteries. Scand J Thorac Cardiovasc Surg 5: 34–38
von der Emde J (1991) Koronare Herzkrankheit. In Borst HG, Klinner W, Oelert H (eds) Herzchirurgie. Springer Berlin, Heidelberg, New York, pp 339–368
Engel HJ, Torres C, Page HL (1975) Major variations in anatomical origin of the coronary arteries: angiographic observations in 4,250 patients without associated congenital heart disease. Cathet Cardiovasc Diagn 1: 157–169
Engel HJ, Dalichau H (1976) Gefährdung der ektopen Circumflex-Koronararterie beim prostethischen Klappenersatz. Thoraxchirurgie 24: 75–79
Ensslen R, Schwarz F, Thormann J, Schlepper M (1977) Überleben und Kollateralzirkulation bei Koronararterienverschluß. Herz Kreisl 9: 20–24
Esperenca Pina JA (1974) Injection-corrosion-fluorescence in the study of human coronary arterial anastomoses. Acta Anat 90: 481–488
Estes EH, Entman ML, Dixon HB, Hackel DB (1966a) The vascular supply of the left ventricular wall. Amer Heart J 71: 58–67
Estes EH, Dalton FM, Entman ML, Dixon HB, Hackel DB (1966b) The anatomy and blood supply of the papillary muscles of the left ventricle. Amer Heart J 71: 356–362
European Atherosclerosis Study Group (1988): The recognition and management of hyperlipidaemia in adults. Eur Heart J 9: 571–600
Fallavollita JA, Brody AS, Bunnell IL, Kumar K, Canty JM (1994) Fast computed tomography detection of coronary calcification in the diagnosis of coronary artery disease. Circulation 89: 285–290
Farrer-Brown G, Rowles PM (1969) Vascular supply of interventricular septum of human heart. Brit Heart J 31: 727–734
Ferreira AG, Trotter SE, König B, Decourt LV, Fox K, Olsen EGJ (1991) Myocardial bridges: morphological and functional aspects. Brit Heart J 66: 364–367
Flaherty GT, O`Neill MN, Daly KM, Folan-Curran J (2001) True aneurysm of the left ventricle. Clin Anat 14: 363–368
Floyd WL, Young WG, Johnsrude IS (1970) Coronary arterial-left atrial fistula. Amer J Cardiol 25: 716–722
Forssmann WG (1979) Zur Struktur, Topographie und Nomenklatur der Cotronararterien. Verh Anat Ges 73: 259–264
Frauenfelder M, Schneider J (1982) Muskelbrücken des Ramus interventricularis anterior der A. coronaria sinistra. Pathologe 3: 247–251
Frink RJ, Achor RWP, Brown AL, Kincaid OW, Brandenburg RO (1970) Significance of calcification of the coronary arteries. Amer J Cardiol 26: 241–247
Frink RJ, James TN (1973) Normal blood supply to the human His bundle and proximal bundle branches. Circulation 47: 8–18
Fulton WFM (1965) Coronary arteries. Charles C Thomas, Springfield, Ill, pp 49–75
Futami C, Tanuma K, Tanuma Y, Saito T (2002) Three-dimensional localizations of the proximal part of the coronary arteries and the arterial blood supply of the conducting system in normal human hearts. Rad Surg Anat (in press).

Gates ARC, Huang CLH, Crowley JJ, Gresham A, Shapiro LM, Carpenter TM, Hall LD (1994) Magnetic resonance imaging planes for the 3-dimensional characterisation of human coronary arteries. J Anat 185: 335–346
Gazetopoulos N, Ioannidis PJ, Marselos A, Kelekis D, Lolas C, Avgoustakis D, Tountas C (1976) Length of main left coronary artery in relation to atherosclerosis of its branches. Brit Heart J 38: 180–185
Geiringer R (1951) The mural coronary artery. Amer Heart J 41: 359–368
Gensini GG, da Costa BCB (1969) The coronary collateral circulation in living man. Amer J Cardiol 24: 393–400
Gianelly R, Mugler F, Harrison DC (1968) Prinzmetal's variant of angina pectoris with only slight coronary atherosclerosis. Calif Med 108: 129–132
Giraldo AA, Higgins MJ (1988) Laboratory methods in the study of coronary atherosclerosis. Pathol Ann 23: 217–236
Gotsman MS, Lewis BS, Bakst A (1973) Obstruction of the left main coronary artery- the artery of sudden death. SA Med J 47:641–644
Gould SE (ed) (1953) Pathology of the heart. CC Thomas, Springfield Ill., pp 549–555
Gould KL, Mullani N, Kirkeeide R (1986) Coronary circulation. In Phelps ME, Mazziotta JC, Schelbert HR (eds) Positron emission tomography and autoradiography. Raven Press, New York. pp.113–147
Gould KL (1988) Identifying and measuring severity of coronary artery stenosis. Circulation 78: 237–245
Gouley BA (1950) Anomalous left coronary artery arising from the pulmonary artery (adult type) Amer Heart J 36: 630–637
Greenberg MA, Fish BG, Spindola-Franco H (1989) Congenital anomalies of the coronary arteries. Radiol Clin North Amer 27: 1127–1146
Gregg DE, Fisher LC (1963) Blood supply to the heart. In Field J, Magoun HW (eds) Handbook of Physiology. Amer Physiol Soc, Washington DC, pp 1517–1584
Griggs DM, Tchokoev VV, Chi Chen C (1972) Transmural differences in ventricular tissue substrate levels due to coronary constriction. Amer J Physiol 222: 705–709
Grondin CM, Dyrda I, Pasternac A, Campeau L, Bourassa MG, Lesperanca J (1974) Discrepancies between cine-angiographic and post-mortem findings in patients with coronary artery disease and recent myocardial revascularization. Circulation 49: 703–708
Gross L (1921) The blood supply to the heart. Hoeber New York, pp 11–25
Gross L, Kugel MA (1934) The arterial blood vascular distribution to the left and right ventricles of the human heart. Amer Heart J 9: 165–177
Grottum P, Svinland A, Walloe K (1983) Localization of atherosclerotic lesions in the bifurcation of the main left coronary artery. Atherosclerosis 47: 55–62
Haberl R, Knez A, Becker A, Becker C; Maaß A, Brüning R, Reiser M, Steinbeck G (1998) Stellenwert der Kalkbestimmung mit Elektronenstrahltomographie bei koronarer Herzkrankheit. Radiologe 38: 999–1005
Hackensellner HA (1954) Koronaranomalien unter 1000 auslesefrei untersuchten Herzen. Anat Anz 101: 123–130
Hackensellner HA (1955) Über akzessorische, von der Arteria pulmonalis abgehende Herzgefäße und ihre Bedeutung für das Verständnis der formalen Genese des Ursprunges einer oder beider Coronararterien von der Lungenschlagader. Frankf Z Path 66: 463–470
Hackensellner HA (1956) Über einige typische Varietäten, Anomalien und Missbildungen des Herzens und der herznahen Gefäße unter 1234 Obduktionen. Acta Morph Hung 6: 403–419
Hadziselimovic H (1982) Blood vessels of the human heart. Thieme Leipzig, pp 14–100
Hallman GL, Coley DA, Singer DB (1966) Congenital anomalies of the coronary arteries: anatomy, pathology and surgical treatment. Surgery 59: 133–144
Halpern MH (1954) Extracardiac anastomoses of the coronary arteries in the human newborn. Anat Rec 118: 306–307
Halpern MH (1955) The sinuatrial node of the rat heart. Anat Rec 123: 425–436
Halpern MH (1957) The dual blood supply of the rat heart. Amer J Anat 101: 1–16
Halpern MH, May MM (1958) Phylogenetic study of the extracardiac arteries to the heart. Amer J Anat 102: 469–480

Harken DE, Black H, Dickson JF, Wilson HE (1955) De-epicardialization: a simple effective surgical treatment for angina pectoris. Circulation 12: 955–962
Harlan BJ, Starr A, Harwin FM (1983) Manual der Herzchirurgie. Springer Berlin, Heidelberg, New York, pp 105–157
Haupt HM, Hutchins GM, Moore GW (1983) Right ventricular infarction: role of the moderator band artery in determining infarct size. Circulation 67: 1268–1272
He B, Tan Y, Cheng M, Cui Y (1991) The surgical anatomy of the sinuatrial node. Surg Radiol Anat 13: 123–128
Hecht HS, Aroesty JM, Morkin E, LaRala PJ, Paulin S (1975) Role of the coronary collateral circulation in the preservation of left ventricular function. Radiol 114: 305–313
Helfant RH, Vokonas PS, Gorlin R (1971) Functional importance of human coronary collateral circulation. N Eng J Med 284: 1277–1281
Helfant RH, Gorlin R (1972) The coronary collateral circulation. Ann Intern Med 77: 995–997
Hirakow R (1983) Development of the cardiac blood vessels in staged human embryos. Acta Anat 115: 220–230
Hoganson G, McPherson, Piper P, Gilbert EF (1983) Single coronary artery arising anomalously from the pulmonary trunk. Arch Pathol Lab Med 107: 199–201
Horiuchi T, Abe T, Tanaka S, Koyamada K (1971) Congenital coronary arteriovenous fistulas. Ann Thorac Surg 11: 102–112
Hort W, Kalbfleisch H (1975) Koronarversorgungstypen des menschlichen Herzens. Herz-Kreisl 8: 385–391
Hudson CL, Moritz AR, Wearn JT (1932) The extracadiac anastomoses of the coronary arteries. J Exp Med 56: 919–925
Hutchins GM, Kessler HA, Moore GW (1988) Development of the coronary arteries in the embryonic human heart. Circulation 77: 1250–1257
Hutchinson MCE (1978) A study of the atrial arteries in man. J Anat 125: 39–54
Irino S, Ono T, Shimohara Y (1982) Microvascular architecture of the rabbit ventricular walls; a scanning electron microscopic study on corrosion casts. Scanning Electron Microsc IV: 1785–1792
Ishikawa T, Brandt P (1985) Anomalous origin of the left main coronary artery from the right anterior aortic sinus. Amer J Cardiol 55: 770–778
Isner JM, Roberts WC (1978) Right ventricular infarction complicating left ventricular infarction secondary to coronary heart disease. Amer J Cardiol 42: 885–894
Isner JM, Shen EM, Martin ET, Fortin RV (1984) Sudden unexpected death as a result of anomalous origin of the right coronary artery from the left sinus of Valsalva. Amer J Med 76: 155–158
James TN, Burch GE (1958) Blood supply of the human interventricular septum. Circulation 17: 391–396
James TN (1961) Anatomy of the coronary arteries. Hoeber New York, pp 88–91
James TN (1965) Anatomy of the coronary arteries in health and disease. Circulation 32:1020–1033
James TN (1977) The sinus node. Amer J Cardiol 40: 965–986
James TN (1980) Neural control of the heart in health and disease. Advances Int Med 26: 317–345
Johnson RA, Blake TM (1968) Vasa vasorum of the heart. Amer Heart J 76: 79–89
Johnston LL, Pohost GM (1994) Nuclear cardiology. In Schlant RC, AlexanderRW (eds) Hurst's The Heart, 8th ed., McGraw-Hill, New York, pp 2281-
Jokl E, McClellan JT, Ross GD (1962) Congenital anomaly of left coronary artery in the young athlete. J Amer Med Ass (JAMA) 182: 174–175
Jokl E, McClellan JT, Williams WC, Gouze FJ, Bartholomew RD (1966) Congenital anomaly of left coronary artery in young athletes. Cardiologia 49: 253–258
Jones PH, Patsch J, Gotto AM (1994) Coronary heart disease. Basic considerations. The biochemistry of blood lipid regulation and the assessment of lipid abnormalities. In Schlant RC, Alexander RW (eds) Hurst´s The heart. 8th ed. McGraw-Hill, New York, pp 973–987
Kafrouni G, Khan AH, Wolfsen JL (1981) Single right coronary artery: clinical and angiographic findings with surgical management. Ann Thorac Surg 32: 80–84.
Kalbfleisch H, Schreck M, von Wedel HR, Hort W (1974) Untersuchungen über die Versorgungsgebiete der Koronararterien . Verh Dtsch Ges Path 58: 526–527
Kalbfleisch H (1975) Eine Methode zur postmortalen Größenbestimmung der Versorgungsgebiete einzelner Herzkrankarterien. Z Kardiol 64: 987–994

Kalbfleisch H, Hort W (1976) Die Verteilungsmuster der Koronararterien (Versorgungstypen) des menschlichen Herzens. Dtsch Med Wschr 101: 1092–1097
Kalbfleisch H, Hort W (1977) Quantitative study on the size of coronary artery supplying areas post-mortem. Amer Heart J 94: 183–188
Kaltenbach M, Becker HJ (1974) Die selektive Koronarangiographie. Dtsch Ärztebl 71: 137–148
Kaltenbach M, Spahn F (1975) Koronarographische Nomenklatur und Typologie der Koronararterien des Menschen. Z Kardiol 64: 193–202
Karadede A, Tacar O, Akkus M (2001) Replacement of circumflex branch of left coronary artery originating from the right aortic sinus. Surg Radiol Anat 23: 281–283
Katz J, Milliken MC, Stray-Gundersen J, Buja LM, Parkey RW, Mitchell JH, Peshock RM (1988) Estimation of human myocardial mass with MR imaging. Radiology 169: 4495–4498
Kaul S, Wismer GL, Brady TJ, Johnston DL, Weyman AE, Okada RD, Dinsmore RE (1986) Measurement of normal left heart dimensions using optimally oriented MR images. Amer J Roentgenol (AJR) 146: 75–79
Kaul S (1989) A look at 15 years of planar thallium-201 imaging. Amer Heart J 118: 581–601
Keith A, Flack MW (1906) The auriculo-ventricular bundle of the human heart. Lancet 2: 359–364
Kennel AJ, Titus JL (1972) The vasculature of the human sinus node. Mayo Clin Proc 47: 557–561
Keren A, Tzivoni D, Granot C, Yalon S, Stern S (1982) Origin of both coronary arteries from the same sinus of Valsalva. Cathet Cardiovasc Diagn 8: 191–199
Kessler W, Laub G, Achenbach S, Ropers D, Moshage W, Daniel WG (1999) Coronary arteries: MR angiography with fast contrast-enhanced three dimensional breath-hold imaging initial experience. Radiology 210: 566–572
Kimbiris D, Iskandrian AS, Segal BL, Bemis CE (1978) Anomalous aortic origin of the coronary arteries. Circulation 58: 606–615
Kitzman DW, Edwards WD (1990) Age-related changes in the anatomy of the normal human heart. J Gerontology 45: M 33–39
Knez A. Becker C, Becker A, Leber A, Haberl H, Reiser M, Steinbeck G (2000) Bildgebende Verfahren in der Diagnostik des Herzens. Radiologe 40: 103–110
Köhler F, Bless H, Pittner PM (1981) Röntgenanatomische Studien an postmortalen menschlichen Koronarostien. Fortschr Röntgenstr 134: 476–482
Koischwitz D, Harder T, Schuppan U, Thurn P (1982) Das Koronararterienaneurysma. Fortschr Röntgenstr 1136: 359–365
Koizumi M, Kawai K, Honma S, Kodama K (2000) Anatomical study of a left single coronary artery with special reference to the various distribution pattern of bilateral coronary arteries. Ann Anat 182: 549–557
Koops B, Kerber RE, Wexler L, Greene RA (1973) Congenital coronary artery anomalies. J Amer Med Ass (JAMA) 226: 1425–1429
Kugel MA (1927) Anatomical studies on the coronary arteries and their branches. Amer Heart J 3: 260–270
Kyriakidis M, Vyssoulis G, Barbetseas J, Toutouzas P (1988) A clinical angiographic study of the arterial blood supply to the sinus node. Chest 94: 1954–1057
Lackner K, Landwehr P, Thurn P (1988) Conventional computerized tomography of the heart. Int J Cardiac Imaging 3: 39–45
Lametschwandtner A, Lametschwandtner U, Weiger T (1990) Scanning electron microscopy of vascular corrosion casts – technique and applications: updated review. Scanning Microscopy 4: 889–941
Langer L (1880) Die Foramina Thebesii im Herzen des Menschen. Sitzungsber Math Naturwiss Classe, Kaiserl Akad d Wiss (Wien) 82 (Juniheft): 25–39
Leary T, Wearn JT (1930) Two cases of complete occlusion of both coronary orifices. Amer Heart J 5: 412–423
Levin DC, Baltaxe HA (1972) Angiographic demonstration of important anatomic variations of the posterior descending coronary artery. Amer J Roentgenol 116: 41–49
Levin DC, Fellows KE, Abrams HL (1978) Hemodynamically significant primary anomalies of the coronary arteries. Circulation 58: 25–34
Lewis CM, Dagenais GR, Friesinger GC (1970) Coronary arteriographic appearances in patients with left bundle-branch block. Circulation 41: 299–307

Liberthson RR, Sagar K, Berkoben JP, Weintraub RM, Levine FH (1979) Congenital coronary arteriovenous fistula. Circulation 59: 849–854
Liberthson RR (1989) Case 22–1989, presentation of case. New England J Med 320: 1475–1483
Licata RH (1955) The developmental basis of the blood supply to the human heart. Anat Rec 121: 330–331
Licata RH (1956) A continuation study of the development of the blood supply of the human heart. Part II: the deep or the intramural circulation. Anat Rec 124: 326 (Abstract)
Lipsett J, Cohle SD, Berry PJ, Russell G, Byard RW (1994) Anomalous coronary arteries: a multicenter pediatric autopsy study. Pediatr Pathol 14: 287–300
Loop FD (1981) Risks in diagnosis and therapy of left main coronary atherosclerosis. Chest 79: 386–387
von Lüdinghausen M (1975) Das Verteilungsmuster der Koronararterien und ihr Einbau in das Myokard. Dtsch Med Wschr 47: 2448–2451
von Lüdinghausen M, Valeske PG (1978) Morphologie und Topik der Verkalkungsherde der Herzkranzarterien und ihre Bedeutung für die arterielle Versorgung des Reizleitungssystems. Herz 3: 325–335
von Lüdinghausen M, Valeske PG (1979) Stenosierende Verkalkungen der Koronarien und Variabilität der Blutversorgung des Reizleitungssystems. Verh Anat Ges 73: 761–763
von Lüdinghausen M, Ohmachi N, Besch S, Mettenleiter A (1995) Atrial veins of human heart. Clin Anat 8: 169–189
von Lüdinghausen M, Ohmachi N (2001) Right superior septal artery with normal right coronary and ectopic early aortic origin: a contribution to the vascular supply of the interventricular septum of the human heart. Clin Anat 14: 312–319
Macchi C, Orlandini SZ, Orlandini GE (1994) An anatomical study of the healthy human heart by echocardiography with special reference to physiological valvular regurgitation. Ann Anat 176: 81–86
Mahowald JM, Blieden LC, Coe JI, Edwards JE (1986) Ectopic origin of a coronary artery from the aorta. Chest 89: 668–672
Margolis JR, Chen JT, Kong Y, Peter RH, Behar VS, Kisslo JA (1980) The diagnostic and prognostic significance of coronary artery calcification. Radiology 137: 609–616
Marshall RC, Tillisch JH, Phelps ME (1983) Identification and differentiation of resting myocardial ischemia and infarction in man with positron computed tomography. Circulation 67: 766–778
Maseri A, Mimmo R, Chierchia S, Marchesi C, Pesola A, L'Abbate A (1975) Coronary artery spasm as a cause of acute myocardial ischemia in man. Chest 68: 625–633
Masuda Y, Naito S, Uda T, Yamada Z, Morooka N, Watanabe S, Inagaki Y (1991) Clinical importance of coronary calcification detected by CT. Acta Cardiol 46: 51–56
Mautner GC, Mautner SL, Froehlich J, Feuerstein IM, Proschan MA, Robers WC, Doppman JL (1994) Coronary artery calcification: assessment with electron beam CT and histomorphometric correlation. Radiology 192: 619–623
May AM (1960) Surgical anatomy of the coronary arteries . Dis Chest 38: 645–657
McAlpine WA (1975) Heart and coronary arteries. An anatomical atlas for clinical diagnosis, radiological investigation, and surgical treatment. Springer, Berlin, Heidelberg, New York, pp 133–209
Menke DM, Waller BF, Pless JE (1985) Hypoplastic coronary arteries and high takeoff position of the right coronary ostium. Chest 88: 299–301
van Mierop LHS (1979) Morphological development of the heart. In Berne RM, Sperelakis N, Geiger SR (eds) Handbook of physiology, Vol 1. The cardiovascular system. Amer Physiol Soc Bethesda, Maryland, pp 1–28
Mizukami S, Tanaka S, Moriya M (1992) Branches of the internal thoracic artery of rats supplying the heart, the sinus region and the brachicephalic veins. Okajamas Folia Anat Jpn 68: 393–408
Mizukami S, Tanaka S, Nakatani T (1996) An autopsy case of an extracoronary artery presenting evidence for the dualism of blood supply to the human heart. Acta Anat Nippon 71: 551–556
Moberg A (1967a) Anatomical and functional aspects of extracardial anastomoses to the coronary arteries. Pathol Microbiol (Basel) 30: 689–694
Moberg A (1967b) Anastomoses between extracardiac vessels and coronary arteries. Acta Radiol Diagn 6: 177–192
Moberg A (1968) Anastomoses between extracardiac vessels and coronary arteries. Acta Med Scand, Suppl. 485: 1–26

Mochizuki S (1933) Vv. cordis. In Adachi B (ed) Das Venensystem der Japaner. Kenkyusha, Tokyo, pp 41–64
Moir TW, Driscol TE, Eckstein RW (1964) Thebesian drainage in the left heart of the dog. Circul Res 14: 245–249
Montenegro MR, Eggen DA (1968) Topography of atherosclerosis in the coronary arteries. Lab Invest 18: 586–593
Moore KL (1988) The developing human, clinically oriented embryology. 4th ed. Saunders, Philadelphia, pp 286–321
Möttönen M (1969) Myocardial infarction and coronary atherosclerosis in Finland. Acta Pathol Microbiol Scand 205:1–89
Müller MF, Fleisch M, Kroeker R, Chatterjee T, Meier B, Vock P (1997) Proximal coronary artery stenosis: three dimensional MRI with fat saturation and navigator echo. J Magn Reson Imaging 7: 644–651
Murphy DA, Roy DL, Sohal M, Chandler BM (1978) Anomalous origin of left main coronary artery from anterior sinus of Valsalva with myocardial infarction. J Thorac Cardiovasc Surg 75: 282–285
Naidoo DP, Sewdarsen M, Vythilingum S, Mitha AS (1988) Anomalous origin of the right coronary artery from the left sinus of Valsalva. S Afr Med J 74: 420–421
National Institute of Health (1993) – conference statement: triglyceride, high density lipoprotein, and coronary heart disease. J Amer Med Ass (JAMA) 269: 505–510
de Nef JJE, Varghese PJ, Losekoot G (1971) Congenital coronary artery fistula; analysis of 17 cases. Brit Heart J 33: 857–862
Nerantzis CE, Koutsaftis PN (1998) Variant of the left coronary artery with an unusual orgin and course. Clin Anat 11: 357–371
Nerantzis CE, Marianu SK (2000) Ectopic high origin of both coronary arteries from the left aortic wall: anatomic and post-mortem angiographic findings. Clin Anat 13: 383–386
Ness MJ, McManus BM (1988) Anomalous right coronary artery origin in otherwise unexplained infant death. Arch Pathol Lab Med 112: 626–629
Noltenius H, Haake A, Giersch H, Buchholz M, Raydt HJ (1976) Pathologisch-anatomische Diagnosen bei 70- bis 102jährigen Verstorbenen. Med Klin 71: 2163–2169
Nomina Anatomica (1980) 5th ed., Williams & Wilkins, Baltimore, London
Ogden JA (1970) Congenital anomalies of the coronary arteries. Amer J Cardiol 25: 474–479
Ogden JA, Goodyer AVN (1970) Patterns of distribution of the single coronary artery. Yale J Biol Med 43: 11–21
O'Rahilly R (1971) The timing and sequence of events in human cardiogenesis. Acta anat 79: 70–75
Page HL, Engel HJ; Campbell WB; Thomas CS (1974) Anomalous origin of the left circumflex coronary artery: recognition, angiographic demonstration and clinical significance. Circulation 50: 768–773
Paschal CB, Haacke EM, Adler LP (1993) Three-dimensional MR-imaging of the coronary arteries. J Magn Res Images (JMRI) 3: 491–500
Paulin S (1964) Coronary angiography. Acta Radiol (Diagn) (Stockholm) 233: 44–133
Paulin S (1983) Normal coronary anatomy. In Abrams HL (ed) Coronary arteriography. Little, Brown and Company Boston, pp 127–174
Paulsen S, Vetner M (1973) Anatomical variations of the coronary arteries and origin of blood supply to sino-auricular and atrio-ventricular nodes determined on the basis of post-mortem coronary angiography. Acta Path Microbiol Scand A 81: 784–790
Pennell DJ, Bogren HG, Keegan J, Firmin DN, Underwood SR (1996) Assessment of coronary artery stenosis by magnetic resonance imaging. Heart 75: 127–133
Penther P, Boschat J, Morin JF, Blanc JJ, Granatelli D (1977) The length of the left main coronary artery. Amer Heart J 94: 705–709
Pepler WJ, Meyer BJ (1960) Interarterial coronary anastomoses and coronary arterial pattern. Circulation 22: 14–24
Petelenz T (1965a) Extracoronary blood supply of the sinuatrial (Keith-Flack's) node. Cardiologia 47: 57–67
Petelenz T (1965b) Extracoronary shunts with coronary arteries in man. Cardiologia 47: 323–336
Pettigrew RI (1994) Magnetic resonance imagingof the heart and great vessels. In Schlant RC, Alexander RW (eds) Hurst´s The heart, 8th ed., McGraw-Hill, New York, pp 2339–2360

Piegger J, Kovacs P, Ambach E (2001) Extremely high origin of the right coronary artery from the ascending aorta. Clin Anat 14: 369–372

Piek JJ, Becker AE (1989) Postmortem quantification of collateral vessels in the human heart. Amer J Cardiovasc Pathol 2: 301–308

Pitt B (1959) Interarterial coronary anastomoses: occurrence in normal hearts and in certain pathologic conditions. Circulation 20: 816–822

Polacek P (1961) Relation of myocardial bridges and loops on the coronary arteries to coronary artery occlusion. Amer Heart J 61: 44–52

Positron Emission Tomography Panel, Council on Scientific Affairs (1988) Application of positron emission tomography in the heart. J Amer Med Ass (JAMA) 259: 2438–2445

Prinzmetal M, Simkin B, Bergman HC, Kruger HE (1947) Studies on the coronary circulation. II. The collateral circulation of the normal human heart by coronary perfusion with radioactive erythrocytes and glass spheres. Amer Heart J 33: 420–442

Prinzmetal M, Kennamer R, Merliss R, Wada T, Bor N (1959) A variant form of angina pectoris: preliminary report. Amer J Med 27: 375–388

Prinzmetal M, Ekmekci A, Kennamer R, Kwoczynski JK, Shubin H, Toyoshima H (1960) Variant form of angina pectoris: previously undelineared syndrome. J Amer Med Ass (JAMA) 174: 1794–1800

Provenza DV, Scherlis S (1959) Demonstration of muscle sphincters as a capillary component in the human heart. Circulation 20: 35–41

Pyke D, Symons C (1951) Calcification of the aortic valve and of the coronary arteries. Brit Heart J 13: 355–363

Ranganathan N, Burch GE (1969) Gross morphology and arterial supply of the papillary muscles of the left ventricle of man. Amer Heart J 77: 506–516

Rath S, Har-Zahav Y, Battler A, Agranat O, Schneeweiss A, Rabinowitz B, Neufeld HN (1986) Frequency and clinical significance of anomalous origin of septal perforator coronary artery. Amer J Cardiol 58: 657–658

Rathor AL, Gooch AS, Maranhao V (1973) Survival through conus artery collateralization in severe coronary heart disease. Chest 63: 840–843

Rauber-Kopsch (1987) Anatomie des Menschen, Bd. 2, (ed. Leonhardt H), Thieme Stuttgart, New York, pp 30–60

Reig-Vilallonga J, Loncan-Vidal MP, Domenech-Mateu (1987) Coronary arterial anastomoses. Anat Anz (Jena) 164: 1–12

Reig J, Jimenez C, Jornet A, Palet J, Petit M (1988/89) Arterial supply of the atrio-ventricular bundle and its right branch by the first diagonal artery: a dual vascularization. Acta Morphol Neerl-Scand 26: 1–7

Reig J, Jornet A, Petit M (1993a) Segmentary analysis of the coronary artery distribution in the left ventricle. Surg Radiol Anat 15: 91–97

Reig J, Jornet A, Petit M (1993b) Patterns of the coronary artery irrigation in the left ventricle. Surg Radiol Anat 15: 309–314

Reig J, Alberti N, Petit M (2000) Arterial vascularisation of the human moderator band. Clin Anat 13: 244–250

Reig J, Petit M (2001) Perfusion of myocardial segments of the right ventricle: role of the left coronary artery in infarction of the right ventricle. Clin Anat 14: 142–148

Reiss P, Becker AE (1981) Dominance of the right coronary artery: what does it mean ? Anat Clin 2: 369–372

Riede UN, Müntefering H, Drexler H, Schaefer HE (1993) Kardiovaskuläres System. In: Riede UN, Schaefer HE (eds) Allgemeine und spezielle Pathologie, 3. Aufl. Thieme, Stuttgart, pp 436–524

Rigby WFC, Graboys TB (1981) Current concepts and managment of the preexcitation syndromes. J Cardiovasc Med 6: 277–293

Riopel DA (1993) The heart. In Stevensen RE, Hall JG, Goodman RM (eds) Human malformations and related anomalies, Vol II. Oxford University Press, New York, Oxford, pp 237–253

Rissanen V (1970) Double contrast technique for post-mortem coronary angiography. Lab Invest 23: 517–520

Robb JS (1965) Comparative basic cardiology. Grune & Stratton, New York, London, pp 123–140

Roberts WC, Morrow AG (1969) Compression of anomalous left circumflex coronary arteries by prosthetic valve fixation rings. J Thoracic Cardiovasc Surg 57: 834–838

Rodriguez FL, Robbins SL, Banasiewicz M (1961) The descending septal artery in human, porcine, equine, ovine, bovine, and canine hearts. Amer Heart J 62: 247–259
Rose AG (1978) Multiple coronary arterioventricular fistulae. Circulation 58: 178–180
Sandstede J, Pabst T, Beer M, Kellner M, Kenn W, Neubauer S, Hahn D (1998) Three-dimensional coronary MR angiography with the navigator technique: imaging of the proximal course of anomalous coronary arteries. Fortschr Röntgenstr 169: 432–436
Sanudo-Tejero JR, Rodriguez-Baeza A, Domenech-Mateu JM (1989) The anomalous origin of the circumflex artery (Ramus circumflexus) in the right aortic sinus. Anat Anz (Jena) 169: 295–299
Schaffler GJ, Groell R, Peichel KH, Rienmüller R (2000) Imaging the coronary venous drainage system using electron-beam CT. Surg Radiol Anat 22: 35–39
Schaper W, Schaper J, Xhonneux R, Vandesteene R (1969) The morphology of intercoronary anastomoses in chronic coronary artery occlusion. Cardiovasc Research 3: 315–323
Schaper W (1974) Zur Entstehung eines Kollateralkreislaufes bei Koronararterien-verschlüssen. Dtsch Med Wschr 99: 2299–2302
Schelbert HR (1994) Positron emission tomography. In Schlant RC, Alexander RW (efs) Hurst's The Heart, 8th ed., McGraw-Hill, New York, pp 2361–2380
Schildt P, Stanton E, Beck CS (1943) Communications between the coronary arteries produced by the application of inflammatory agents to the surface of the heart. Ann Surg 118: 34–45
Schlesinger MJ (1938) An injection plus dissection study of coronary artery occlusions and anastomoses. Amer Heart J 15: 528–568
Schlesinger MJ (1940a) Significant variations in the anatomic pattern on the coronary arteries. Blood Heart Circulation 13: 61–72
Schlesinger MJ (1940b) Relation of anatomic pattern to pathologic conditions of the coronary arteries. Arch Path 30: 403–415
Schlesinger MJ, Zoll PM (1941) Incidence and localization of coronary artery occlusion. Arch Path 32: 178–188
Schlesinger MJ, Zoll PM, Wessler S (1949) The conus artery: a third coronary artery. Amer Heart J 38: 823–836
Schoenmakers J (1963) Verteilungstypen der Koronararterien. In (Bargmann W, Doerr W eds.) Das Herz des Menschen, Bd. 2, Thieme, Stuttgart, pp 735–792
Schwartz JN, Kong Y, Hackel DB, Bartel AG (1975) Comparison of angiographic and post-mortem finds in patients with coronary artery disease. Amer J Cardiol 36: 174–178
Scott DH (1948) Aneurysm of the coronary arteries. Amer Heart J 36: 403–421
Scotti TM (1977) Heart. In Anderson WAD, Kissane JM (eds) Pathology, 7th ed. Mosby, St. Louis, pp 737–855
SelvesterRH (1982) cited after Reig et al. (1993)
Sharbaugh MAH, White RS (1974) Single coronary artery. J Amer Ass Med 230: 243–246
Silverman ME, Hurst JW (1968) The mitral complex. Amer Heart J 76: 399–418
Sow ML, Ndoye JM, Lo EA (1996) The artery of the atrioventricular node. Surg Radiol Anat18: 183–187
Smith SC, Gorlin R, Herman MV, Taylor WJ, Collins JJ Jr (1972) Myocardial blood flow in man: effect of coronary collateral circulation and coronary by-pass surgery. J Clin Invest 51: 2556–2560
Sobotta (2000) Atlas der Anatomie des Menschen, Bd. 2, 21. Aufl., Putz R, Pabst R (eds), Urban & Fischer München, Jena, pp 76–91
Spalteholz W (1924) Die Arterien der Herzwand, anatomische Untersuchungen am Menschen- und Tierherzen. Hirzel, Stuttgart
Spring DJ, Thomsen JH (1973) Severe atherosclerosis in the single coronary artery. Report of a previously undescribed pattern. Amer J Cardiol 31: 662–665
Stolte M, Weis P, Prestele H (1977) Die koronare Muskelbrücke des Ramus descendens anterior. Virchows Arch Path Anat 375: 23–36
Stroobandt R, Riessens J, Suy R (1974) Unstable angina pectoris in single coronary artery. Acta Cardiol 29: 469
Tanaka S, Lee HY, Mizukami S, Nakatani T, Chung IH (1998) Posterior sinus node artery and accessory atrioventricular node artery arising by a common origin. Clin Anat 11: 106–111
Taylor JR (1980) The descending septal artery. Arch Pathol Lab Med 104: 599–602
Terminologia Anatomica (1998) International Anatomical Terminology, Federative Committee on Anatomical Terminology, Thieme, Stuttgart, pp 78–79

Truex RC, Smythe MQ, Taylor MJ (1967) Reconstruction of the human sinoatrial node. Anat Rec 159: 371–378

Turner K, Navaratnam V (1996) The positions of coronary arterial ostia. Clin Anat 9: 376–380

Vlodaver Z, Neufeld HN, Edwards JE (1972) Pathology of coronary disease. Semin Roentgenol 7: 376–394

Vlodaver Z, Frech R, van Tassel RA, Edwards JE (1973) Correlation of the ante-mortem coronary angiogram and the post-mortem specimen. Circulation 47: 162–169

Vlodaver Z, Neufeld HN, Edwards JE (1975) Coronary arterial variations in the normal heart and in congenital heart disease. Academic Press, New York, San Francisco, London, pp 1–55

Vlodaver Z, Amplatz K, Burchell HB, Edwards JE (1976) Coronary heart disease; clinical angiographic and pathologic profiles. Springer New York, Heidelberg, Berlin, pp 1–158

Voboril Z, Schiebler TH (1969) Über die Entwicklung der Gefäßversorgung des Rattenherzens. Z Anat Entwickl Gesch 129: 24–40

Waller BF (1983) Five coronary ostia: duplicate left anterior descending and right conus arteries. Amer J Cardiol 51: 1562

Waller BF, Schlant RC (1994) Anatomy of the heart. In Schlant RC, Alexander RW (eds) Hurst's The Heart, 8th ed. McGraw-Hill, New York, pp 59–112

Wearn JT, Mettier SR, Klumpp TG, Zschiesche LJ (1933) The nature of the vascular communications between the coronary arteries and the chambers of the heart. Amer Heart J 9: 143–164

Williams PL, Warwick R, Dyson M, Bannister LH (eds.) (1989) In Gray´s Anatomy, 37th British ed., Churchill Livingston, London, pp 727–732

WilsonPM (1965) An unusual variation of the coronary arteries. Anat Anz 116: 299–302

Wright NL (1970) Coronary artery steal due to an anomalous left coronary artery originating from the pulmonary artery. J Thorac Cardiovasc Surg 59: 461–467

Subject Index

GPSR Compliance

The European Union's (EU) General Product Safety Regulation (GPSR) is a set of rules that requires consumer products to be safe and our obligations to ensure this.

If you have any concerns about our products, you can contact us on ProductSafety@springernature.com

In case Publisher is established outside the EU, the EU authorized representative is:

Springer Nature Customer Service Center GmbH
Europaplatz 3
69115 Heidelberg, Germany

Batch number: 10704713

Printed by Printforce, the Netherlands